A *Healthy* HOME FLOCK

Quarto.com

First Published in 2027 by Cool Springs Press, an imprint of The Quarto Group,
100 Cummings Center, Suite 265-D, Beverly, MA 01915, USA.
T (978) 282-9590 F (978) 283-2742

EEA Representation, WTS Tax d.o.o.,
Žanova ulica 3, 4000 Kranj, Slovenia.
www.wts-tax.si

31 30 29 28 27 1 2 3 4 5

ISBN: 978-1-57715-757-1

Digital edition published in 2027
eISBN: 978-1-57715-758-8

Library of Congress Cataloging-in-Publication Data available

Design and Page Layout: Samantha J. Bednarek, samanthabednarek.com
Photography: Nikki Husted, except Four Blessings Photography on pages 4, 8, 9, 10, 26, 43 (top), 55, 66, 74, 136, 157 (bottom), 184, and 201; Jackie Chenelli on page 81; Brandie Fogal on page 84; Ashley Reese on page 85; iStock on pages 83 (left), 86, and 137 (top left); Jenifer Thatcher on page 90, 110; Katelyn Anderson on page 92; Samantha Gillis on page 102 (top right); Marlee Quintasket on page 103 (top left); Marianne West on page 103 (middle left); Amy Hunnicutt on page 103 (bottom left); Christine Ellis on page 103 (top right); Rebecca Low on page 103 (bottom right); Veronica Stirnus on page 105 (middle); Tianna Weatherbee on page 111; Shutterstock on pages 113 and 116; Rachel Fouasnon on page 118; Misty Moss on page 137 (top right); Andrea Newcomb on page 139 (bottom left); Tina Richmond on page 142; Ashley Happ on page 150
Illustration: Ada Keesler (pages 24, 35, and 62–66), Samantha J. Bednarek (stethoscope), and Shutterstock (large chicken illustration)

Printed in Guangdong, China TT072026

A Healthy Home Flock

A Complete Guide to Keeping Backyard Chickens and Chicks

HAPPY + HEALTHY

NIKKI HUSTED
author of *Chicken Keeping Pure & Simple*

Foreword by Dr. Blayne Mozisek, The Poultry Doc

COOL SPRINGS PRESS

CONTENTS

This book is dedicated to my grandmother, Lois Varner, who will tell anyone who will listen how much work proper chicken keeping is and how proud she is of me. Care and compassion start with the matriarch, and she has nurtured an entire family of wonderful humans. That is a legacy worth celebrating.

FOREWORD BY DR. BLAYNE MOZISEK, THE POULTRY DOC

Raising chickens is more than just a hobby; it is a commitment to the well-being of our feathered family members. Whether you are currently holding your very first box of chirping chicks or have been tending to a coop for decades, there is one universal truth in poultry keeping: The learning never stops. Our flocks depend on us for their health, safety, and happiness, and as their caretakers, we have a responsibility to continually strive for deeper knowledge, no matter our level of experience. That is exactly why I, as a board-certified poultry veterinarian, am so thrilled to introduce this book by Nikki Husted, known to so many of us as @PurelyChickens. She has become a vital voice in the poultry community, and I find this book to be an excellent resource that serves the entire spectrum of enthusiasts—from the absolute novice to the most seasoned veteran.

Finding reliable advice today often feels like an uphill battle. A simple search can lead keepers down a rabbit hole of conflicting opinions and anecdotal evidence that lacks a solid foundation. At The Poultry Doc, my team and I see the fallout of this confusion daily, spending hours untangling internet myths that unintentionally endanger the welfare of flocks.

Nikki's work stands apart because it is rooted in two essential pillars: facts and science. In these pages, she shares incredible insight into the most important aspects of poultry keeping, as well as the seemingly mundane daily tasks that are so often overlooked. She understands that every detail—from the design of a roosting bar to the management of the chicken run—is critical to raising a happy, healthy flock. By focusing on evidence-based husbandry, Nikki provides a road map that cuts through the noise and centers on what actually works.

I know you will find this book to be a valuable addition to your library. It is a guide that empowers you to make informed decisions for your birds, ensuring they live their best lives under your care. I could not be more excited to recommend this book, and I am confident that the wisdom within these chapters will benefit you and your flock for many years to come.

The Poultry Doc
Dr. Blayne Mozisek

INTRODUCTION

MY ROOSTER, BIG MIKE, sent my husband to the hospital. Actually, two hospitals in one weekend. We were attempting to trim Mike's spurs, but while trying to catch him, my husband had to maneuver over a board. That split-second mistake gave Mike time to spur him in the leg. Within hours, my husband was in agony and couldn't walk. After many tests, it was determined that he had a severe infection in a tendon in his leg. It took a whole week and a lot of antibiotics before he could walk again. You'll never believe it, but Big Mike is still around, protecting the flock to this day.

Why am I telling you this? I want you to know who I am as a chicken keeper before you dive into this book. I do not view my flock as disposable animals. I do not keep chickens only when they serve a purpose. I see my chickens as pets first, eggs second. I care for them like I would my dog or cat. They go to the vet if needed. I keep their coop clean. I thank them for their eggs, name them, give them treats, and I do not dispose of them when they don't do what I want. My chickens are more than just poultry; they are family.

My chickens are more like pets than livestock. Cruella is a family favorite, seen here perching on my shoulder.

Big Mike is a great rooster. Day to day, he goes about his business and protects his ladies. In that one instance, he felt threatened and vulnerable, cornered and unable to reach his flock. If I were him, I would have done the same. The injury wouldn't have been so bad if we had been more diligent about trimming his spurs! So Mike stays, and now we only trim his spurs at night, when he is sleeping and easier to catch.

Now that we got that out of the way, rest assured, I won't judge you if your chickens are animals that must serve a purpose to stay on the farm. I view chicken keeping more as a hobby, and it is a passion for me! But that isn't the case for everyone. If you keep chickens to feed your family and earn money from egg sales, you might have a different viewpoint. No matter your perspective, know that I do not take my job as a chicken educator lightly. I put my heart and mind into researching what is best for both pets and livestock.

A healthy chicken flock requires minimal effort. Beyond feeding and watering, they are mostly self-sufficient. There are fewer losses in a healthy flock, and they lay more eggs! An unhealthy flock constantly deals with issues. They are stressed, so they don't lay eggs as often or at all. They lose more feathers, pick on each other more, and struggle to fight off bacteria and viruses.

In this book, I will cover all aspects of maintaining a happy, healthy flock. I will cover preparation, troubleshooting, routines, and management. You'll learn about the most common and some not-so-common ailments your chickens may face, as well as the best ways to assist them through these problems.

Chickens can be the easiest animals in the world. They can also be difficult if you're constantly troubleshooting. Even if your chickens aren't "pets," you can save time and money by being prepared. Whether that means creating an environment that prevents catastrophic issues or learning to address common problems as they arise, maintaining a healthy flock is essential for anyone keeping chickens.

My coops have changed year after year, but this one is my favorite. The green makes my heart happy.

It's always better to be prepared, whether that means having the tools to get the work done or knowing what to expect.

1

SETTING UP FOR *Success*

Eleanor Roosevelt said, "It takes as much energy to wish as it does to plan." Planning has saved me energy in chicken keeping. When starting a new venture, preparing for various scenarios is wise, though chickens can find themselves in unpredictable situations. I've often thanked myself for a stocked first-aid kit or winter coop supplies bought in advance. Sadly, many chickens are discarded when owners realize how much work is involved. While you can't prepare for everything, having core supplies saves time and heartache. In this chapter, I'll help set you up for success!

ARE YOU READY?

HOLD ON! Before you go into the farm supply store, fall in love with the fluffy chicks, and bring them home on a whim . . . let's think this through. Trust me, I know that's easier said than done. Chicken keeping deserves some planning. Even though it would be disappointing to go home empty-handed, I think it might be worse to have to rehome a flock of chickens later. I'm not here to talk you out of it, but I don't want you to be surprised either. Let me share a few things to consider before you fill that box with chicks.

Restrictions

Local laws and regulations can be one of the biggest surprises for new chicken keepers. I have heard troubling stories about families who had to get rid of their flock because of restrictions imposed by their township. It is always best to check with your local zoning department, HOA (Homeowner's Association) counsel, or city or township website for clarification. Regulations can be extreme or nonexistent, depending on where you live.

COMMON RULES

Cities and towns may have zoning rules that determine whether chickens are permitted in that area. The rules may depend on your property size and zoning classification. If chickens are allowed, you may need a permit or be limited in how many you can keep. Some cities restrict whether you can have a rooster in your flock (even though hens can be louder!). Purchasing female chickens does not guarantee they will be female. Hatcheries cannot guarantee 100 percent accuracy in sexing, so if your city doesn't allow roosters, it's wise to have a plan for any roosters that end up in your flock by mistake. Rules may require your coop and run to be a certain distance from property lines, neighbors' homes, wells, or other structures. There may also be sanitation requirements to keep rodents and other pests at bay.

Finally, if you plan to sell eggs, check with your local department of agriculture and health department to ensure you are following their guidelines.

Emotional Considerations

What you can handle emotionally is deeply personal. I was surprised to become so emotionally attached to my chickens. There's nothing wrong with running a farm to support your family without getting emotionally attached to your chickens. There's also nothing wrong with treating them as pets. If that's you, consider what that means for your backyard chicken flock. Emotional attachment may mean spending extra money to keep them safe, happy, and healthy long after they stop laying eggs.

Chickens get sick and injured. Most illnesses and injuries are relatively simple to handle, but others take a strong stomach. Consider what you'll do if one of your chickens gets a wound that flies find before you do. Picking out maggots from a chicken wound is not something you may have expected, but it must be done to save the chicken's life. I have dealt with spotting a hawk tearing apart my favorite hen, flushing out wounds, having mites crawl on me while I treat the flock, and nursing sick hens back to life. Can you handle these things? For most of us, it is a learned skill, but others may decide this isn't for them.

Time, Money, and Labor

Bringing animals into your life requires a commitment of extra time, money, and labor. There are many ways to manage a chicken coop, and the approach you choose will depend on your preferences and resources. I have found that a basic, low-cost setup can work well, but it typically demands more hands-on labor and time. On the other hand, investing a little extra money up front can mean less daily effort in the long run. Finding a balance that works for you may take some trial and error. It took me years to figure out what worked best for my flock and me.

One thing no one seems to expect is growth. If you build a coop for six hens, you may find yourself building a bigger one sooner rather than later. While six hens might lay the perfect number of eggs for a small family, egg production will decline after a few years. Most people end up adding chicks to the flock every now and then to keep the eggs coming. Other reasons you might want more chickens include beautiful egg colors, different breeds, providing eggs for family and friends, or the irresistible urge to purchase those cute chicks.

My Silkies and Bantams love their smaller coop, and I love that it is easy to maintain.

IMPORTANT COMPONENTS OF A CHICKEN SETUP

THE HOME BASE FOR CHICKENS IS THEIR COOP. They need to feel safe in their coop. You can purchase boxed coops that can be screwed together in an afternoon, but they aren't the safest option. Even larger coops may need adjustments once you realize they don't work for your chicken-keeping style. Starting with a good setup will save you time, effort, and possibly heartache.

The Chicken Coop

The purpose of a chicken coop is to keep your flock safe at night, provide a home base for sleeping, and offer a comfortable place to lay eggs. Some coops are elaborate, while others are simple. Chickens don't need you to fuss

I am a sucker for a cute chicken coop sign! I think my chickens like them too.

Be sure to include a ramp in your setup if your roosts are higher than a few feet. Larger birds can easily be injured by jumping off roosts.

over them. Half the things I do to my coop are more for me than for them. Believe me, my chickens couldn't care less about the window curtains or wallpaper. That said, there are some things to consider when deciding on your coop.

COOP PLACEMENT

Where you place your coop is almost as important as the design. Chickens need a dry, shaded space. If you live in a climate with hot summers, I suggest finding a spot with plenty of shade. If your summers stay cool, you don't need to worry about shade as much. Choose a dry location because a wet, soggy environment breeds bacteria and increases the risk of foot infections (not to mention it's smelly!).

SIZE

Your chicken coop needs to fit your future flock, not just the six you started with! It's wise to plan ahead for a larger group. Even if you never add another chicken, your flock will appreciate the extra space! I also suggest keeping yourself in mind when building a coop. Think about how you can easily access each area of the coop. You do not want to be hunched over or leaning far into the coop while cleaning.

A common rule of thumb is to provide 3 square feet per chicken inside the coop. For example, a 4- by 8-foot coop (1.2 by 2.4 m), which is 32 square feet (2.9 square m), could house about ten chickens. This is a great starting point, but the number can vary based on ventilation, roosting space, and litter management.

Take my coops, for instance. My main coop is 8 by 8 feet (2.4 by 2.4 m), which is 64 square feet (5.8 square m). This should accommodate twenty-one chickens, yet it currently houses forty! How do I get away with this? My litter management system involves removing most, if not all, of the droppings each morning, leaving the coop ammonia-free for the next night. I have ample ventilation, including windows, solar-powered fans, a roof ridge vent, and side vents. I dust my coop regularly and installed extra roosting bars to ensure everyone has a spot. There is even unused space on the roosts every night.

SLEEPING ARRANGEMENTS

Most chicken breeds prefer to sleep on the highest spot in the coop. Like wild birds, they like to perch on a roost. But unlike wild birds, chickens are round and fat, so I highly suggest using flat roosting bars to give them a flat area to distribute their weight. Chickens that roost on smaller bars often develop foot deformities or bumblefoot (see page 84) from uneven pressure. Flat roosts also allow them to cover their feet in winter, which helps prevent frostbite. I use pieces of untreated 2 × 4s (5 × 10 cm) positioned with the 4-inch (10 cm) side facing up.

Ruth loves the roosting spot by the window, so she secures her spot before dark!

NESTING

Nesting boxes are an essential addition to your coop. Without them, your chickens will lay eggs wherever they choose, including in the yard, in the weeds, or in hidden corners of the barn. If you don't want to go on a daily egg hunt, nesting boxes are a must!

I prefer 12- by 12-inch (30 by 30 cm) nesting boxes, suitable for larger hens or two hens arguing over a spot. I've tried various bedding materials, like straw, pine shavings, rubber mats, artificial grass, and sand, but nest pads are best because they prevent bedding from kicking out and are easy to replace. A common rule is one nest box per four hens, but there may be unused nest boxes. To encourage hens to use certain boxes, I place fake or real eggs inside, as hens tend to lay eggs in places that others have deemed safe.

Your nests should be used only for laying eggs. Chickens defecate at night, so if they make a habit of sleeping in them, your eggs will be covered in droppings. No one wants a dirty egg, so keep those chickens out! If they are too young to lay eggs, cover the boxes with cardboard or something similar to keep them out. If you can't cover them, take the hen out after dark and place her on the roost. I do this every night until a better habit is formed.

REMOVABLE ROOSTS Make your roosting bars removable with joist hangers. My dad came up with this idea, and I have seen it used everywhere since! This lets you remove them to clean them in the yard. I use a power washer. Dirty roosts lead to foot infections!

PRO TIP

Hens prefer to lay eggs in safe places. Dark, quiet, hidden spots are their favorite! Try adding curtains to the boxes for a more secure feel.

Shhh! Franny is busy laying an egg!

Having this safe space for birds being integrated has been so helpful!

INTEGRATION STATION

Whether you are adding chickens to your flock, separating a mother hen and her chicks, or giving an injured chicken a safe place to heal, an integration station is a wonderful asset to the coop! My old coops didn't have a designated spot, so I had to pull out a dog crate or put the chickens I needed to isolate in my garage. Now I have a convenient spot for them with the flock while they heal or integrate. This can be designed in many ways, but mine is under the shelf. It has doors that hinge and lock. When I'm not using it for chickens, I use it for storage. I do not suggest using this space to separate sick hens, as they may be contagious (more on isolating sick hens on page 37).

A CASE FOR BIG FEEDERS I prefer feeders that allow multiple chickens to eat at once and hold 50 pounds (22.7 kg) of feed. I also like feeders that let the chickens still see their surroundings, so if a higher-ranking chicken is approaching, the chicken can move out of the way and avoid getting hurt. My favorite waterers are nipple or cup waterers because they stay cleaner and there is less spilling. Nipple waterers can be heated in the winter to prevent freezing. Keep food off the ground and water out of the sunny areas of the run.

I can't imagine that rolling in dirt feels good, but chickens absolutely love it!

DUST BATH

You wouldn't build a house without a shower, and you shouldn't build a coop without a dust bath! It may seem counterproductive, but chickens stay clean by bathing in dirt! It is a natural behavior, and you may even notice chicks trying to bathe in their bedding or a bowl of feed if given the chance. Dust bathing is important for parasite control (see page 71), oil balance on the feathers, temperature regulation, stress relief, and bonding with the flock. I offer my flock a place to dust bathe all year-round. A covered run helps by keeping the area dry, allowing the chickens to dust bathe.

ENRICHMENT

Your flock doesn't need to watch the latest popular show, but they do need enrichment in their run to keep them entertained. From bullying to feather picking, chickens can get feisty. Providing enrichment will save you from trouble later on. You will see many ideas online for toys, bells, and mirrors. While these things may be helpful at times, chickens tend to get used to them and then ignore them altogether. I find it more useful to add branches, ladders, and roosts for climbing and perching. I also add swings and treat dispensers for added enrichment.

DUST BATH HOW-TO To make a dust bath, simply combine dry dirt with coarse-grit sand to keep it loose. I also add a lime product available in the Resources section at the back of the book. My dust bath area is the hangout spot. The chickens congregate here, especially when the afternoon sun hits it.

To get the chickens used to the swing, I screwed wooden bowls onto the ends and added treats to encourage them to give it a try.

LITTER MANAGEMENT

IF ANYTHING WILL MAKE YOU QUIT chicken keeping, it's poop. Peck my toes? Fine. Destroy my flower beds? Okay. Get into dangerous situations daily? We'll deal with it. But a stinky, gross coop will make anyone want to throw in the towel. That's why litter management isn't just about cleanliness. It's also about making chicken keeping sustainable and enjoyable. A well-managed coop shouldn't smell horrible. Of course, there are days I get behind, and it's less pleasant, but for the most part I'm able to maintain a coop that's enjoyable to be around. Find a method that works for you and your schedule.

Hemp bedding is soft, low dust, and super light!

Bedding Options for the Coop

There are many types of bedding to choose from. Here are some pros and cons for each.

- **Straw** is easy to find and inexpensive, especially in rural areas. It was probably the bedding most of our great-grandparents used! Unfortunately, it isn't the most absorbent and can harbor mold when wet. It can also attract pests such as mice and mites. Straw needs to be changed frequently. Don't confuse straw with hay, which is similar but used as livestock feed.
- **Pine shavings** are inexpensive and easy to find. They work reasonably well in deep-litter systems. There are a few downsides, including low absorbency, potential respiratory irritation, and dust. I think the biggest downside is that they are made from pine trees, so they aren't environmentally friendly.
- **Sand** can be sifted like a litter box to remove droppings. I used to use it in my coop and loved it . . . until my chickens got mites, over and over again. I had to shovel out all that heavy sand to clean properly and then find a place to dump it. It can be extremely dusty, especially inside a closed coop in winter. My chickens were regularly getting clogged nostrils! I do love using sand in my run, where there is open air. If you use sand in the coop, avoid play sand and look for something with coarse grit and minimal dust.
- **Hemp** is a newer bedding option on the market. To me, it is by far the superior bedding for chicken keeping. I use it on my coop floor and in all of my chick brooders. Depending on the brand, it can be low dust. Hemp is highly absorbent, making it ideal for controlling odors and moisture. It is an environmentally friendly plant that doesn't come from trees and is easily composted, soft on their feet, and easy on their respiratory systems. The downside is that it can be expensive. This is usually balanced by the fact that you don't have to change it as often.
- **Leaf litter** is something you may use if it's readily available. Bonus: It's free! It must be dried before using it in the coop, or it will cause mold to grow and can introduce pests to your coop if you aren't careful. It isn't absorbent, so it isn't great at keeping odors down, but it is wonderful in the compost.
- **Shredded paper** seems great but can be disappointing. Although it's free, eco-friendly, and low dust, it clumps when wet, sticks to their feet, needs frequent changes, and doesn't absorb moisture well.

Coop Setups

The coop style you choose will depend on your flock size, climate, personal preferences, and how you want to balance your time and effort for coop maintenance. The good news is that with a little planning, you'll be able to find a system that works for you. In this section, I'll outline the most common coop styles, along with the pros and cons of each, so you can find a system that fits your lifestyle.

SH!T SHELF

The sh!t shelf method involves building a shelf under roosting bars to collect droppings each night. These are made of plywood covered with vinyl flooring stapled in place. I clean it in the morning and add moisture-absorbing products, preferring those listed in the Resources section. This method minimizes ammonia, takes as much time to clean as to collect eggs, and keeps the coop fly-free.

Even with the shelf, the coop needs bedding on the floor. We call it bedding, but chickens don't sleep in it like other farm animals. Why do we need it if they don't sleep on it? Occasionally, they poop on the floor, track mud, or break eggs. Bedding absorbs mess and cushions their feet when jumping off roosts. I clean it with a sifting shovel as needed, but not often, since they rarely poop outside the shelf.

DEEP LITTER

The deep litter method is my least favorite chicken coop setup, though many do love it. I find it hard to control smell and dislike seeing gnats or flies around. I prefer many small cleanups over less frequent heavy ones. Deep litter depends on proper composting on the coop floor, with droppings layered with bedding that composts in place. It can help in cold climates if started early enough to generate heat. To maintain, start with about 4 inches (10 cm) of bedding, add more weekly, and turn it occasionally to promote microbial breakdown. Cleanup is needed two or three times a year, then restart the process.

Morning chores make for a clean coop all day long.

Janice is admiring her clean coop! Nothing like an orderly coop with fresh bedding.

With a sifting tool, you can easily remove droppings and feathers from sand or bedding.

OCCASIONAL SIFTING

Some coops are easily managed with the occasional sifting of droppings. If the coop is small and holds just a few chickens, this method might make the most sense. I use this method in my Bantam coop because they don't have a shelf under their ladder-style roosts. They are also smaller chickens with tiny droppings that dry out quickly, unlike my larger flock, whose droppings don't. I use hemp bedding on the floor of their coop and I sift the droppings once a week (more often if the flies are bad). I add more bedding as needed.

RAISED WIRE FLOOR

Wire floors under the roosts with air space below is better suited for warmer climates, so I couldn't get away with it in Ohio! Droppings fall through the wire onto the ground below, where they can be scooped away as needed, or the coop can be moved to a different spot. One downside to this method is that predators could get into the coop if it is not secured properly. Even if they can't get in, they can still get under the wire, which can stress the flock. If you go with this method, consider closing any gaps to prevent this from happening.

HOOP HOUSE

A hoop house is usually made from PVC, metal piping, or sometimes wood, with chicken wire or hardware cloth secured over it. It is sometimes covered with a tarp to keep the chickens dry. Often, there is no enclosed coop, and the chickens sleep in the hoop house with an attached roost. This setup is better suited to hot climates without cold winters. Hoop houses can also be made mobile.

Chicken Runs

Some farmers or backyard chicken keepers have a barn or coop for their chickens to use as needed and then let them free-range during the day. If you don't have the space for that, or if you have too many predators (see page 58) to let them free-range, your chickens will appreciate a safe, dry place to be when they aren't in their coop. A chicken run provides a space for chickens to engage in natural behaviors like eating, dust bathing, and socializing. The area is usually attached to the chicken coop and provides extra space for them. Your flock will use the run to eat, drink, and hang out during the day. Even free-range flocks benefit from a run during vacations, predator threats, or bad weather. Whether fenced or enclosed, it keeps chickens safe and active.

The run is where you place the food and water for the flock. Keeping them out of the coop helps keep the area cleaner and pest-free. The food must be kept dry because if it gets wet, it will get moldy and become toxic to the flock. Place the waterers where they don't get direct sun to prevent algae growth. It's best to keep both off the ground to avoid attracting pests such as ants and mice. Make sure both food and water are easily accessible to you and the chickens.

I ended up adding more Silkies and Bantams than I had planned, so I had to add a run extension!

My main flock has had three additions to their run. Luckily, my dad knows how to build and rebuild a great roof using scrap materials!

COVERED VS. UNCOVERED

I always suggest putting a roof on your chicken run if you have the means. Chickens thrive in dry spaces, and an uncovered run is exposed to the elements. One rainstorm can leave your run wet and smelly, harboring bacteria. Chickens walking in these conditions can develop respiratory illness, coccidiosis, or bumblefoot. A covering also protects them from the sun. Just like any animal, too much sun can cause overheating, dehydration, or, believe it or not, sunburn! A roof on the run also provides extra protection against predators and critters like mice and rats.

SIZE

The size of your run, unlike the coop, depends mainly on your flock's size. Because the run is open and airy, ventilation won't be as critical. The most important thing is giving the flock ample space to move around and interact safely. Cramped spaces lead to more fighting and bullying. I suggest 10 to 15 square feet (1 to 1.4 square m) per chicken in the run. If you can make it bigger, even better! When designing your run, consider headroom for yourself. The run will need maintenance, and you'll want to be able to access the chickens as needed. Being able to stand upright is a great perk for any setup.

ATTACHED RUN

My preferred way to create a chicken run is to attach it to the henhouse. It is typically built with a sturdy wooden frame and wrapped in hardware cloth. I advise against using chicken wire, as some predators can still break through. Adding a roof is ideal to keep the space usable during rain and snow. I also suggest making it as spacious as possible. Usually, 10 to 15 square feet (1 to 1.4 square m) per chicken is enough room for them to live comfortably.

FENCING

Some chicken keepers prefer to fence in an area to keep their chickens contained rather than building a larger enclosed run. If you have a large coop or barn for them to use during bad weather, this approach might work well. I suggest covering the top with bird netting to deter overhead predators. You could also place the fencing around trees that provide safety and shade.

CHICKEN TRACTOR

If you have a lot of land and some muscle, a chicken tractor might be a good option for you. By building a chicken tractor, you make the coop and run mobile. You can lift and roll it around your land every few days. This gives the flock new foraging ground and makes cleaning easier for you. If you have machinery to help you move the tractor, you may be able to build one for a larger flock, but chicken tractors tend to work best for small flocks.

My flock has a protected run but also has access to a fenced-in area for a bit of freedom.

COMBINATION COOP

A chicken run that includes roosting bars for sleeping but lacks a separate enclosed space to serve as a coop house is often called a combination coop. It can be a cost-effective way to contain your flock while providing open-air space for chickens in a hot climate. However, it is not ideal for most climates, as the chickens need an enclosure to get out of the wind and cold. It can also attract nocturnal predators because the chickens aren't hidden at night.

BEDDING OPTIONS FOR THE RUN

In my opinion, there are fewer bedding or substrate options that work in the run than in the coop. This is because the run is exposed to more chicken traffic and weather. It needs to stay dry to keep chickens healthy and be easy to clean. In my experience, it is best to use some sort of substrate, as the ground itself will just give the manure a solid place to pack down, making it difficult to clean and irritating to their feet. Here are some of the options:

- **Sand** is my preferred substrate for a chicken run under a roof, especially when mostly dry. It requires weekly cleaning using a sifting shovel to remove manure, feathers, and debris. I spend about 15 minutes a week, and it keeps the run clean year-round. Proper chicken sand prevents impacted crops and dust inhalation, so choose sand made for chickens or coarse-grit all-purpose sand.
- **Wood chips** are an inexpensive, sometimes free, option for chicken keepers and make good compost. Let chickens mix and turn the mulch, then remove it and add fresh mulch at least twice a year, or more often for many chickens. Drawbacks include heaviness; incomplete removal of droppings; and potential harm from certain woods, such as dyed mulch, cedar, and bark, which can harbor mold spores.

You should see small pebbles in the sand, and it should be dry and not compact when you squeeze it.

- **Straw** is an alternative some chicken keepers use. It is similar to wood chips but requires more frequent cleaning because it stays wet and smells. It's not absorbent and can become moldy if left wet by rain or manure. I only use straw when there's heavy snow, as it helps chickens keep their feet warm while outside.
- **Yard litter** such as leaves and pine needles can be used as a more sustainable option. It isn't the most convenient because you have to collect and replace it often. Its maintenance is similar to that of straw or wood chips. Cleaning and turning will depend on how many chickens you have in the space. Be watchful for mold, which is toxic to chickens. It is best to dry the litter before adding it to the run.

MANURE MANAGEMENT

YOUR FLOCK WILL PRODUCE more than just eggs—you'll also be able to collect manure! Okay, I know I made that sound more exciting than it is, but chicken manure is so valuable that people will pay for it! It can be an excellent addition to your garden or property, but if left unmanaged, it can become a major problem. It may attract pests, produce odors, or pose health concerns for your family and flock. As a chicken keeper, you have a few options for managing litter.

Composting

The most valuable way to use chicken manure is to add it to your compost. Chicken manure is considered "hot," meaning it is high in nitrogen and needs to be composted longer than other types of manure. Once composted, it is rich in nitrogen, phosphorus, potassium, and organic matter, making it a great addition to your garden. If left raw, the manure could burn your garden plants and harbor pathogens such as *E. coli* and Salmonella. Here is how I like to compost this gift from my chickens:

1. Using a multibin system, collect droppings (nitrogen) and carbon-rich materials such as leaves, hemp bedding, straw, and other browns. Aim for a 2:1 C:N ratio (two parts carbon to one part nitrogen).
2. Build your pile over time and turn it often to allow for proper aeration.
3. Once the pile reaches about 3 cubic feet (1 cubic m), it should be large enough to retain heat. Wet the pile and use a compost thermometer to ensure it stays at 130°F to 150°F (54.4°C to 65.6°C) for 3 days to kill harmful pathogens.
4. Repeat the heating process by pushing the middle of the pile to the edges, bringing the edges to the middle, wetting it again, and letting it reach the correct temperature.
5. Allow your pile to cool, then cure for about 60 days. The compost is ready to add to the garden when it is dark brown and crumbly.

Aging

If composting is too complicated and you have a lot of space, aging might be your answer! While aging chicken manure isn't as fast as composting, it can still get the job done with minimal effort. Make a pile of chicken manure with bedding and other browns. Keep it dry and let it break down on its own. This can take up to a year. I have multiple spots on my property for aging. You can also add it to your garden at the end of the summer and mix it in when spring begins. This is more of a "set it and forget it" method, and with my busy lifestyle, I appreciate that!

Composting chicken manure can be a real benefit to your garden and land.

Piles like this work well for aging droppings and bedding, especially when you don't have time to tend a compost bin.

Bagging and Disposing

If you don't have the means to compost or age your manure, you can dispose of it in the trash. I suggest using compostable trash bags to offset the environmental impact of adding manure to the landfill. It is also recommended that you double-bag the manure with two bags to protect sanitation workers. While this is an option, it isn't the most cost-effective or environmentally efficient.

Donation

Have you ever visited a local coffee shop and seen bags of used coffee grounds for gardeners to take? You can do the same with manure! You may be surprised by how many people in your community would welcome nutrient-rich chicken poop as a compost additive. Once you build a base of interested people, you might just be known as the poop person in your community. There are worse things!

Vermicomposting

If you are into critters, vermicomposting uses worms to turn chicken manure into rich, nourishing castings called vermicompost. This method stabilizes the nitrogen-rich chicken manure. It works well in bins or beds, including those integrated into a compost system. The material needs to be well-aged and high in carbon, such as bedding, to prevent ammonia from harming the worms. With the help of the worms, your garden will enjoy dark, crumbly castings that enrich your soil.

Black Soldier Fly Larvae Bioconverting

Black soldier fly larvae (BSFL) are an excellent food source for chickens. They are sold dried by almost every chicken-supply company! You can harvest them at home and give your chickens a nutrient-rich, fresh, and sustainable treat. Using their manure to feed the BSFL creates a closed-loop system you can feel good about. It does take some research and practice, but once you have a system in place, it can save you money and get rid of that manure!

WEATHER

YOUR CLIMATE INFLUENCES flock management. Chickens handle cold better than heat, but both pose challenges. Keep breeds suitable for your climate. In hot areas, choose heat-hardy breeds like Leghorns, Rhode Island Reds, and Easter Eggers, which have large combs, small bodies, and lighter feathers. Cold climates need cold-hardy breeds like Orpingtons, Marans, and Brahmas, with smaller combs and larger bodies. Specific weather tips are found in chapter 4.

When you build stability and strength into your basic setup, you dramatically reduce the risk of crisis. Setting up for success doesn't guarantee it. Unfortunately, no matter how hard we try, illness and injury are inevitable parts of chicken keeping. Birds living outside are at greater risk of pathogen exposure, predator attacks, parasites, and more. In the next chapter, we'll cover preparing for and responding to illness and injury. You'll learn what to have on hand in case something happens, how to spot warning signs, and how to help chickens recover and thrive.

Being prepared for problems helps you keep your flock happy and healthy!

2

PREPARING FOR Injuries and Illness

As soon as you decide to become a chicken keeper, you accept responsibility for caring for them when they are injured or sick. Chickens are masters at hiding weakness, doing so to protect themselves from predators that target weak flocks. As you get to know your chickens' normal behaviors, you'll recognize early warning signs of illness or injury. I have been known to drive a long distance to have my chicken seen by a specialized vet when I am dealing with issues I can't treat myself. Unfortunately, there are relatively few veterinarians who specialize in poultry. Many common ailments and minor injuries can be handled at home if you are well prepared. With the right mindset, tools, and knowledge, you will be more likely to save a chicken in need. In this chapter, we'll cover recognizing signs of illness and injury, what to include in your first aid kit, and the critical steps to take in an emergency.

PREVENTION

THE FIRST LINE OF DEFENSE is prevention. You'll save money, time, and heartache by preventing illness and injury before they occur. By planning ahead and adopting good habits, you lower the risk of disease and ensure your chickens have the safe environment they need to thrive. Prevention goes beyond proper nutrition and coop setup. It also includes their social environment, predator exposure, cleanliness, and biosecurity. Below are some of the culprits that can undermine your flock's success.

Nutrition

Most chicken feed brands make it easy to choose the right feed for your chickens based on age and laying status. Chicken nutrition is well studied, and most companies offer a nutritionally complete feed that supports your flock. Steer clear of homemade blends, foods made for other species, and excessive treats. Make sure your flock has access to chicken feed, even if they free-range, and avoid spoilage or mold. I will go into more detail about nutrition in chapter 3.

This chicken is having a refreshing drink after laying an egg.

Automatic doors are great when they work! Don't forget to check the batteries and connection often.

Hydration

Water is essential for digestion, thermoregulation, egg laying, and overall health. Waterers should be kept clean and full daily to prevent dehydration, poor egg production, vent gleet (a type of yeast infection), and reduced nutrient absorption. Keep waterers out of the sun to prevent algae and bacterial growth, clean them weekly with soap and water, keep them off the ground to avoid contamination, and prevent them from freezing in winter.

Coop Setup

Every coop will be different from the next, and believe it or not, your chickens will adjust to your chosen style. However, there are a few tips to keep in mind. Roosting bars shouldn't be too high because even with the best ramp, your chickens will occasionally jump down from the roost, which can cause leg injuries. Choose automatic doors with chicken sensors so no one gets smooshed by a closing door before bed! Be sure to provide enough nesting boxes to reduce squabbling and pecking. Use flat, wide roosting bars so they can distribute their weight evenly, avoid bumblefoot, and keep their feet covered in the winter. Provide ample ventilation inside the coop and avoid drafts directly on the chickens.

Coop Maintenance

Dirty, damp, and poorly ventilated coops breed pathogens and parasites. Regular cleaning of bedding, roosts, and nesting boxes helps prevent this. Flies, mites, lice, mice, and other pests are less likely to invade a clean coop. Good ventilation helps keep your chickens from getting respiratory illness, and quality building materials keep weather and predators out.

A separate pen is the perfect spot to let chickens get used to each other.

This hen is looking out through the safety of hardware cloth . . . probably hoping for treats!

Overcrowding

Too many birds in one space can lead to stress, squabbles, feather picking, rapid spread of pathogens and parasites, ventilation problems, and more. Overcrowding is one of the biggest causes of illness and injury in the flock. Ensure your birds have sufficient space and room to expand when you inevitably decide to get more chickens.

Integration

Integrating new birds into the flock can be tricky (see page 184). The existing flock isn't usually kind to new members. Injuries can occur if new birds aren't introduced properly. Illness can also be an issue if the newcomers aren't quarantined for a few weeks, treated for illness or parasites, and observed for signs of disease.

Biosecurity

Biosecurity refers to the precautions you take to keep disease and parasites out of your flock. Chickens are highly susceptible to illness when biosecurity practices aren't in place. Examples of good biosecurity include quarantining new birds before bringing them into the flock, keeping mice and wild birds away from your chicken feed and area, and changing your clothes and boots after visiting another farm or flock. More information is available in chapter 4.

ASK THE POULTRY DOC

BIOSECURITY

In the world of poultry medicine, the "closed flock" is the gold standard for biosecurity and disease prevention. In its strictest biological definition, a closed flock is a fortress.

Single-Age Cohort: You start with one group of day-old chicks.

Zero Additions: No new birds are ever added.

Zero Re-Entry: No bird leaves the property and returns.

Impermeable Boundaries: No visitors with poultry, no shared equipment, strict control of rodents and wild birds, and rigorous control of all inputs (feed, bedding, water).

This model is standard for the National Poultry Improvement Plan's (NPIP) primary breeding operations and research facilities because it works. However, for the average backyard keeper, a 100 percent closed flock is statistically improbable. Flocks naturally evolve—predation occurs, hens age out of laying, and owners want new breeds.

While you may not achieve a technically "closed" flock, adopting a closed-flock mindset based on veterinary science is the single most effective biosecurity measure a keeper can implement to prevent disease challenges in their flock.

Predator Protection

Not only can predators cause direct death, but even their presence can cause stress in the flock. Building a secure coop with quality materials and using hardware cloth instead of chicken wire can help keep dangerous predators away. Avoid leaving food scraps lying around, especially at night, and keep brush and debris around the coop to a minimum to avoid attracting pests and predators. I will discuss predator protection in greater detail in chapter 4.

Seasonal Considerations

Being prepared for every weather condition will prevent many issues with your flock. Frostbite, sunburn, heat exhaustion, egg production issues, organ failure, dehydration, bumblefoot, and respiratory illness are just a few problems that can be easily avoided with a bit of weather prep.

Parasite Control

Internal and external parasites can take a toll on your flock. Lice, mites, and worms can cause anemia, malnutrition, decreased egg production, and even death. It is best to conduct regular feather, skin, and fecal tests. We will cover all of this in chapters 6 and 7.

SIGNS OF A SICK CHICKEN

By the looks of Olive's body language, she isn't feeling like herself.

Brenda isn't feeling great. Labored breathing is a symptom of illness.

NO MATTER HOW HARD we work to prevent illness and injury, some things are unavoidable. Recognizing when a chicken is sick or hurt will get easier over time. Unfortunately, chickens are experts at hiding weakness. They do this instinctually for two reasons. First, the flock is more vulnerable to predators if there are sick or injured chickens. Second, other flock members will pick on the weaker ones, so they hide their weakness. Chicken keepers may miss early warning signs and only spot the problem when it is too far progressed for the chicken to mask. As you get to know your flock's movements, posture, interactions, and appetite, you'll get better at spotting early warning signs. Let's look at the key warning signs to watch for and how to distinguish an easy fix from an emergency.

General Signs of Illness

- **Behavior changes:** Normal, healthy chicken behaviors include scratching, foraging, dust bathing, and staying alert to their surroundings. A sick chicken will stop some of these activities and become lethargic and less energetic. They may isolate themselves from the flock, stand in one place, or close their eyes frequently.
- **Loss of appetite:** An ill chicken eats and drinks less. If they turn down their favorite treats or have empty crops at the end of the day, you may be dealing with an illness.
- **Egg production decline:** Viruses and parasites can be present in a hen's eggs. You may notice misshapen eggs, poor-quality eggs with thin shells, reduced egg production, or even no eggs at all. Abnormal eggs are normal to see every once in a while, especially in a young or old hen.
- **General appearance:** To assess a chicken's health, check feathers, skin color, eyes, body, and posture. Dry, brittle, or fluffed feathers suggest problems. Combs and wattles should be pink or red; any pale, white, blue, or purple tones need evaluation. Eyes should be clear and round, without deformation, puffiness, bubbles, or discharge. Unhealthy chickens may lose weight, have a bony breastbone, or look thin. They often stand hunched, with ruffled feathers and droopy wings, tails, and necks.
- **Respiratory symptoms:** Chickens can show respiratory symptoms like coughing, sneezing, runny nose, swollen eyes, eye discharge, or labored breathing when sick with infection. These illnesses spread fast, so watch for these signs!

- **Droppings:** Get to know your flock's droppings. Stools from a sick chicken may be black, contain blood or mucus, or be loose and watery. Check out chapter 6 for images of normal and abnormal droppings.
- **Unnatural movements:** Chickens can develop neurological issues and viral infections that alter their movements. If your chicken has a twisted neck, is walking backward or off balance, is trembling, or has paralysis, you may want to consider reaching out to a vet.

Signs of Injury

- **Visible wounds:** You may be able to see cuts, puncture wounds, or blood on the chicken. If you see the injury, you should also look for inflammation around the wound, swelling, heat, and signs of infection.
- **Lameness:** Limping, unwillingness to bear weight, or dragging may signal a leg injury or a hidden cut under feathers. I once had a chicken limping with a severe wing cut. Inspect carefully and patiently to find injuries.
- **Feather damage:** Missing or broken feathers can mean the chicken had an encounter with a predator or a flock member. Broken feather shafts can bleed and may need to be pulled.
- **Skin issues:** Skin irritation can occur for many reasons. Whether it's parasites or another irritant, unhealthy skin can be uncomfortable. Look for red skin without feathers.
- **Swelling or bruising:** Check legs, toes, and wings for swelling or bruising; bruised chickens usually look greenish. Struggling chickens should be separated and kept low to the ground while healing.
- **Eyes:** Whether it's caused by a tree branch or a peck from a friend, they can get eye injuries that result in swelling, infection, and cloudiness.
- **Head, neck, and spine issues:** Injuries can cause neurological symptoms like imbalance, spinning, or walking backward. Twisted neck or spine, inability to stand, and drooping neck are signs of injury.
- **Beak injuries:** Pecks on hard surfaces, fights, or wall collisions after a fright can cause injuries. Small cracks or breaks are harmless as the beak regrows. Bleeding, large missing parts, or deformations are concerning.

A well-stocked first aid kit will save you from scrambling time and time again!

FIRST AID KIT

IF YOU ARE PREPARED for chicken illness and injury, the likelihood of helping them recover is much higher. The first step is to compile a first aid kit to cover all bases. It's best to have many useful items on hand so you can either handle problems yourself or help the chicken until you can get them to a veterinarian. I have used my first aid kit more times than I can count, and each time I am so grateful it is ready to go. Be mindful of medications that may expire so you can replace them as needed. Let's go over some items you will want to have.

Supplies:

- **Gauze:** Gauze or nonstick pads are used to cover and protect open wounds or things like bumblefoot.
- **Self-adherent wrap:** This resembles a wrap you might use for your own sprained ankle, but it sticks to itself, which is perfect for wrapping chickens without having to use adhesive material that may stick to feathers.
- **Medical tape:** This may be used for extra support when wrapping injuries or creating a splint.

- **Tools:** A good first aid kit will have scissors, pet nail trimmers, tweezers, syringes, scalpels, eyedroppers, a thermometer, a towel, paper towels, and possibly tube-feeding equipment.
- **Gloves:** Disposable gloves will be helpful to keep you and your bird safe.
- **Blood coagulant:** Cornstarch or another blood-clotting agent will come in handy for broken feathers, nails, or beaks.
- **Chicken saddle:** These are made to keep the rooster from damaging the hen during mating, but saddles are also great for covering back injuries or missing feathers.
- **Heat plate:** Even older chickens can benefit from being warmed up now and then! When they are very sick, warmth is essential.
- **Heat lamp:** I don't love using heat lamps long term around chickens unless I need to warm a chick or an adult chicken quickly.
- **Popsicle sticks:** These are useful for creating splints.
- **Scale:** In some cases, the chicken needs to be monitored to make sure they aren't losing weight.
- **Thermometer:** An adult chicken's temperature should be 105°F to 107°F (40.6°C to 41.7°C). A rectal thermometer may be used.

Medications and Supplements:

- **Saline:** This is used to flush wounds or remove debris from the eyes.
- **Chlorhexidine:** This is used for more severe wounds that risk infection. I use 0.5 percent dilution.
- **Antiseptic wound-cleaning spray:** I find that a wound-cleaning spray saves time and stress (for both the chicken and me). A good spray helps flush wounds, prevents infection, and aids in healing.
- **Antibiotic ointment:** This can aid healing. Note that the pain-relief versions are toxic to chickens.
- **Electrolytes:** When a chicken is lethargic, having trouble staying alert, or dealing with heatstroke, electrolytes are used.
- **Probiotics:** These can be helpful for gut issues or illness.
- **Terramycin:** This antibiotic eye ointment is helpful for eye infections and injuries.
- **Amprolium:** This medication is off-label for chickens. It is found in the cattle section of your local farm store and is meant to treat coccidiosis. One brand I have used is Corid.
- **Aspirin:** Aspirin can be used for pain relief.
- **Vitamins:** Vitamins are helpful during times of illness, for replenishing the body after the use of antibiotics or amprolium, or during a vitamin deficiency. There are several brands of chicken-specific vitamins. I prefer liquid vitamins, and I keep a general mix along with a paste of vitamin E with selenium.
- **Epsom salt:** Epsom salt is used for egg binding, bumblefoot soaking, and muscle soreness.
- **Personal lubricant:** You can use this when a chicken is egg-bound (see page 142).
- **Antibiotics:** In the United States, antibiotics for animals require a veterinarian's prescription. This is to prevent medication resistance. It is best to have a veterinarian on hand to help.
- **Natural products:** I use natural products with my chickens as a first line of defense. I keep oregano and thyme essential oils on hand for viral and bacterial illnesses. I also make my own ointment for frostbite, which you can find in chapter 5.

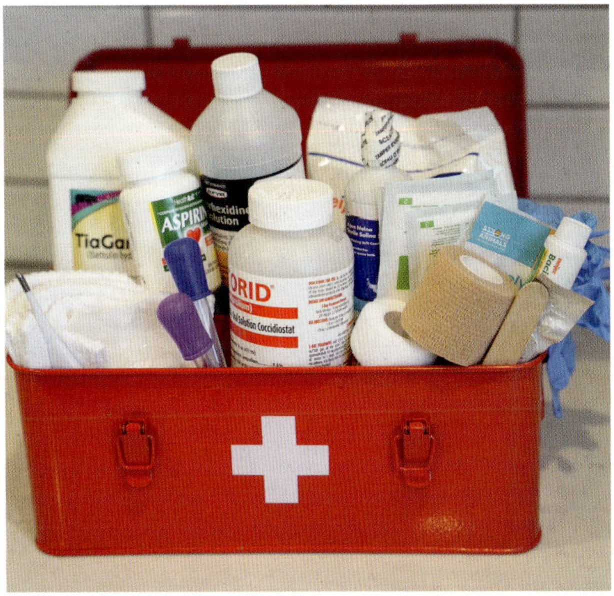

Don't forget to check labels for storage instructions and expiration dates. I like to double-check my first aid supplies yearly.

Tips for Administering

We are all used to taking medication ourselves, but giving medication to a bird can be a daunting task! Their small mouths and throats, the risk of aspiration, and the stress both of you are under all demand extra care. Below are some practical strategies to make the process safer and more effective.

A towel can be used to wrap the chicken before giving medication or dressing wounds. This protects you and them.

1. **Know the Basics**
   - Only administer medications approved for poultry or under the direction of your vet. Unapproved (or off-label) medications can cause issues for you and your chicken.
   - Check the expiration dates, dosage, and your bird's weight.
   - Review the withdrawal times before starting. Many medications have recommended egg and meat withdrawal periods to ensure the medication has cleared the bird's system and you aren't consuming it yourself.
   - Read the labels for possible reactions, regulations, dosage, and administration instructions.
2. **Prepare Tools, Medication, and Space**
   - Gather the essential tools. Items you may need include a needleless syringe for liquids, a soft rubber catheter for forced oral delivery, a dropper, a sterile needle and syringe, a bowl for crushing pills if needed, and an old towel for restraint.
   - Prepare and clean a safe spot. You'll want to be away from the flock in a quiet, clean, and distraction-free space.
3. **Prepare the Chicken**
   - Act calmly around your chicken to reduce stress and flapping. It will be much harder to administer meds to a scared chicken.
   - Use a towel to restrain the chicken's wings. One flap could spill the medication or ruin the dosage. Avoid applying pressure to their chest and airways. Do not hold them upside down; instead, use an underarm cradle or lap hold. Administering meds to a chicken on their back, especially liquids, can cause aspiration.
4. **Medicating Water Source**
   - Dissolve the prescribed dose in the specified amount of water.
   - Observe the flock to be sure they are consuming the liquid.
   - Pay attention to the directions about how often to dump and remix the medicated water. Usually, this must be done daily.
   - Make sure the chickens aren't being underdosed because they aren't drinking the solution.
5. **Oral Medications**
   - Gently restrain the chicken or have a helper do this part for you.
   - Use a syringe or dropper to administer the medication gently. Insert the tip into the side of the beak and allow the bird to swallow before pushing more through.
   - If you hear gurgling or coughing, or see fluid regurgitate, stop administering the medication to prevent aspiration.
6. **Tube Feeding or Gavage Feeding**
   - This should be done under a veterinarian's guidance, especially if the chicken has difficulty swallowing, is lethargic, refuses medication, or has a beak injury. Not all medicines can be tube-fed, as some need slow water absorption.

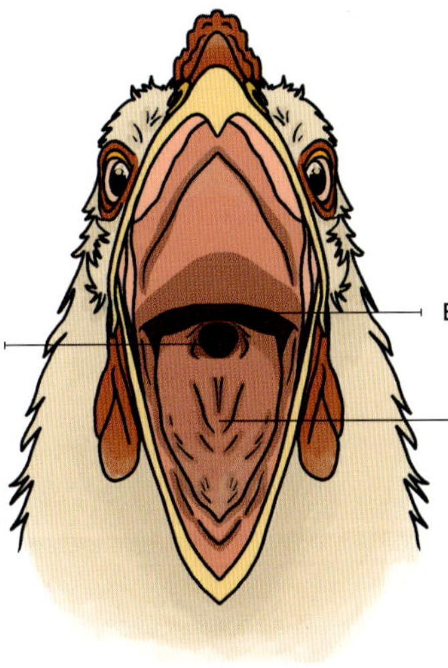

Always administer medication or fluid to the sides of the mouth to avoid liquid entering the trachea. Chickens cannot close their trachea, and if fluids enter, they will aspirate.

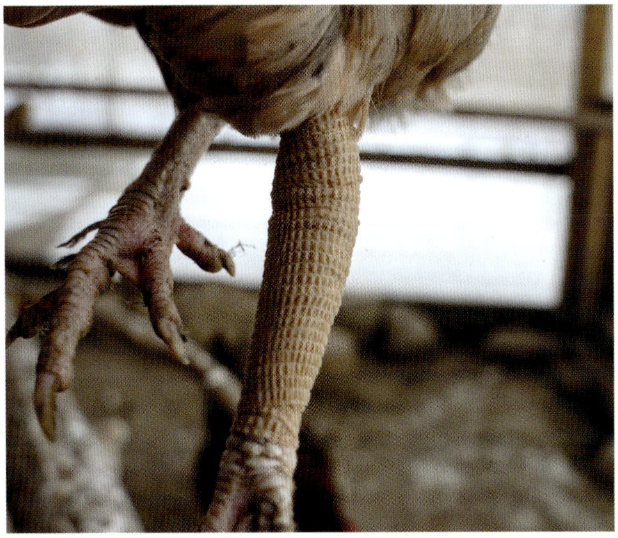

Self-sticking vet wrap works well for wrapping a chicken's wound.

- Tubes can be purchased online or through your vet and are intended for one-time use. They are usually size 8 to 12 French for adult hens.
- Measure the chicken from the beak to the crop and mark the tube at that length. This will prevent the tube from being inserted too far.
- Load a syringe with the exact dose of medication and attach it firmly to the tube.
- Insert the opposite end of the tube into the bird's mouth on the right side. Be careful to aim for the esophagus on the right side, not the trachea, which is in the middle. Stop advancing the tube if you feel resistance.
- Slowly push the medication through the syringe.
- Remove the tube slowly and keep the bird calm and upright until everything settles.

7. **Topical**
   - If you are dealing with a wound or lesion, clean the area first with an antiseptic safe for poultry.
   - Apply the ointment, spray, gel, or medication gently.
   - Use nonstick gauze to cover the wound. Secure the gauze to the chicken with self-adhering vet wrap, and be careful not to cut off circulation.

8. **Eyes**
   - Use a gentle saline solution if you need to flush debris from the eye.
   - Wipe gently with clean, saline-soaked gauze if there is discharge or crust on the eye.
   - Clean your hands well or wear disposable gloves.
   - Gently pull back the bottom eyelid and put in the ointment (¼ inch [6 mm]) or drops.

9. **Injections**
   - Check with your vet before administering medication by needle. They can give you proper instructions or show you in person.
   - You will need a sterile needle, a syringe, medication, an alcohol swab, and a container for safe disposal of the needle.
   - Subcutaneous injections (under the skin) are usually used for vaccinations, fluids, or slow-absorbing medications and given in loose skin at the back of the neck or between the wings.
   - Intramuscular injections (into the muscle) are used when giving antibiotics, vitamins, and some medications. They usually go into the breast muscle.
   - Intravenous injections (into a vein) are used for severe, emergent dehydration or for anesthesia. A vet should do this.

CHICKEN INFIRMARY

Moira needed to spend time away from her flock due to a respiratory illness. After a few days of care she was feeling better.

SETTING UP A PROPER isolated space helps your chicken heal from illness or injury. Separating your chicken allows you to monitor their food and water intake, administer medicine, and keep their environment quiet. Flock members can be brutal to weak ones, so the infirmary improves their chances of recovery. Keeping them separate also prevents other flock members from contracting illnesses and helps stop contagious outbreaks early. Have supplies ready for quick setup.

If the chicken is sick with something contagious, keep the infirmary far from the flock to prevent germs from spreading through air or contact. I recommend at least 40 feet (12 m) away and not upwind from the coop. I use my garage because I can keep it closed to contain germs. The space should be protected from weather. In spring and summer, my garage is warm enough for a chicken to rest. In winter, I bring them to the basement or add heat to the garage. It's easier for them to heal if they are not using energy to stay warm. The space should be quiet, away from noise and traffic. I make my husband park his truck in the driveway (he loves that).

The infirmary should be large enough for the chicken to move around, turn, and stretch, but not so large that it's hard to keep warm. If the enclosure is too big, an injured chicken can further injure itself by jumping and not resting the injury. I don't add a roosting bar in mine because I don't want a weak chicken to fall off or an injured chicken to get any risky ideas.

Here are some items you will need in your infirmary:

- **Flooring:** This should be solid and easy to clean. Wire-bottom cages should be avoided.
- **Bedding:** Choose soft, low-dust bedding, such as hemp. If the chicken has an open wound where bedding could get trapped, old, clean towels are a better choice.
- **Water:** Your chicken will need water at all times. Because the enclosure is small, I prefer a spill-proof drinker. The waterer should also stay clean. A hanging waterer is a great option to accomplish these things.
- **Food:** Make sure your chicken has access to food as needed. Unless the chicken has a beak injury, use a shallow feeder. It should also be easy to clean and stable so it doesn't tip.
- **Heat source:** Most of the time, you won't need a heat source, but it's important to have one in case you need to help the chicken maintain body heat. If you're going to leave it plugged in unattended, make sure it's safe, such as a radiant heater. I use a heat lamp to warm a chicken quickly.
- **Light:** Natural light or lights you can turn on and off with the day cycle are best. Avoid bright lighting that can cause more stress.
- **Ventilation:** Fresh air is essential, but you should avoid drafts, especially in colder weather. The enclosure shouldn't be sealed tight; it should have windows or wire sides.
- **Storage:** I like keeping all of my "sick chicken" supplies in a large lidded container. It is labeled and ready to go when I need it.

EMERGENCY CHECKLIST

A heat lamp warms up my sick chicken quickly! I use my mudroom sink and hang the light from the cabinets above.

WHEN YOU FIND A SICK or injured chicken, every second counts. Chickens hide their weakness until it is no longer doable, and by the time you notice, time is of the essence. Staying calm and completing all the necessary steps will ensure the best outcome for your feathered friend. This checklist will come in handy the next time your chickens need you!

Step 1: Isolate the Chicken

Remove your chicken from the flock immediately, whether sick or injured, to protect the flock and improve recovery chances. Choose a quiet, draft-free area far from the flock. Set up an infirmary with bedding, feeder, waterer, and heat source if needed.

Step 2: Assess the Chicken

Understand the symptoms and dangers. Check the bird's airway and breathing. Signs of distress include open-mouth breathing, panting, wheezing, or gurgling. Open-mouth breathing or panting can mean heat exhaustion, airway obstruction, or respiratory illness. Wheezing indicates poor air quality, infection, obstruction, or mites. Gurgling sounds are wet and bubbly, suggesting fluid buildup or aspiration. Check their temperature; if hot, it may indicate infection. Observe their face coloring: Pink or red is normal; pale or white can signify anemia, frostbite, parasites, stress, liver issues, organ failure, or other illnesses. Blue or purple may show oxygen deprivation or poor circulation but can also be stress-induced, which often normalizes quickly.

Step 3: Regulate the Chicken's Temperature

If the bird is chilled, stiff, or lethargic, warm it before anything else. I've revived stiff, sick chickens by warming them with a heat lamp, heating pads, or blankets. I rotate the bird under the lamp, avoiding burns, and ensure the temperature is safe. I also gently rub their body to stimulate and keep them alert.

For an overheated chicken, it is safer to lower their body temperature gradually. Move them into the shade, use a fan to circulate air around them, and offer water with electrolytes. You can fill a bucket with room-temperature or slightly cooler water and submerge their body for a few minutes at a time. Be careful to avoid their mouth, nose, and ears.

A quick cooldown helps a chicken deal with summer.

Step 4: Hydrate the Chicken

After warming (or cooling, as the case may be) the chicken, hydration is the next most important step. There are several ways to go about this, and the method you choose will depend on how your chicken is functioning. The best-case scenario is that your chicken can drink on their own.

- **Oral hydration:** Oral hydration involves giving chickens water from a bowl or drinker, suitable for mildly dehydrated chickens that can still swallow. Sick or injured chickens benefit from electrolyte or vitamin mixes in their water, which help balance water in cells and prevent swelling or shriveling. Water should be at room or body temperature to avoid shock.
- **Oral-assisted hydration:** If your chicken refuses or can't get liquids, you may need to assist. An eyedropper is best for control. Needle-less syringes work but can be hard to control and may cause aspiration. In a pinch, a metal straw is another option. Offer water from the end by placing it to the beak tip, holding water with a finger. The key is to offer water one drop at a time from the side of the beak. Avoid forcing, squirting, or giving too much at once, as chickens can aspirate easily.
- **Subcutaneous or intravenous fluids:** Fluid is injected under the skin or into a vein. Chickens that are severely dehydrated or unable to take water orally could benefit from this method. While some chicken keepers do this at home, I suggest reaching out to a veterinarian if your chicken has reached this level.

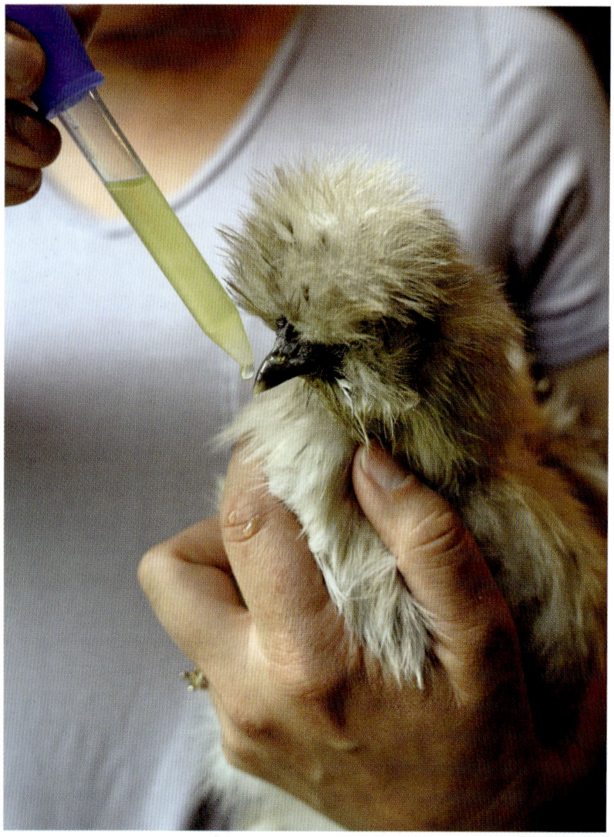

This chicken needed fluids, so I gave her one drop at a time with an eyedropper. Go slow and steady so she doesn't aspirate.

Step 5: Give the Chicken Nutrients

Once a chicken's temperature is regulated and they are hydrated, offer quickly digestible food. Distressed chickens may refuse dry crumbles or pellets, so moisten feed with warm water for a mash, which is more palatable. Scrambled eggs are also nutritious; if they refuse, add yolk to the water or give in a bowl. Note that raw eggs can contain bacteria, so use at your own risk and wash the eggs before cracking.

What you do after these emergency steps depends on the situation and your comfort. If your chicken is still struggling, lethargic, or declining, seek help. It's wise to have a plan for suffering chickens, whether culling (see page 190) or a vet on speed dial.

FINDING A VETERINARIAN

ONE OF THE BIGGEST CHALLENGES for backyard chicken keepers is finding a vet who is well versed in chickens. For centuries, chickens have been viewed as disposable livestock. If one chicken became sick, the farmer or homesteader would cull it and move on. Today, some of us are becoming more attached to our chickens and are willing to pay for veterinary care. Finding that care can be the biggest challenge.

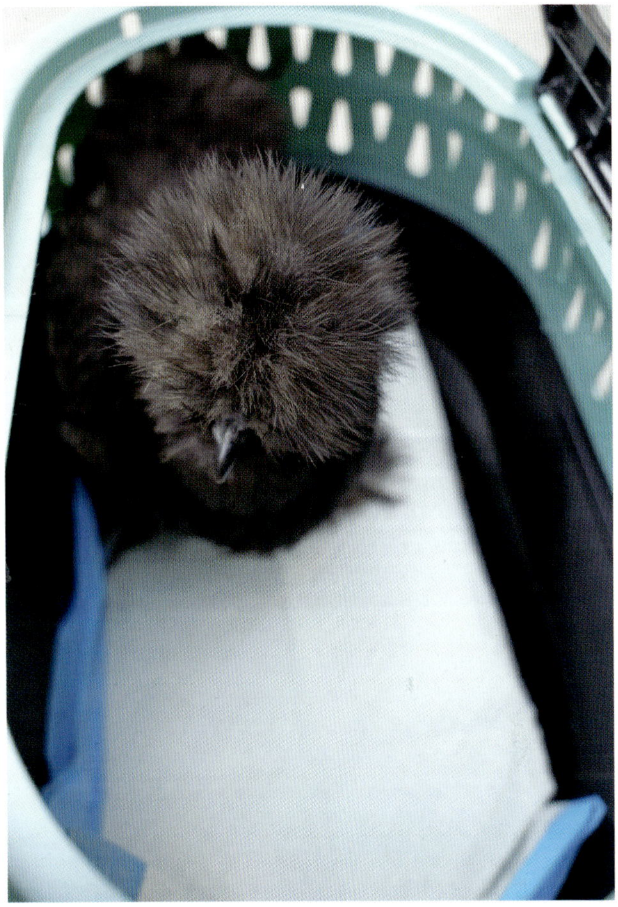

When visiting a vet, use a cat carrier or small crate to keep the chicken contained and limit flapping so they don't hurt themselves.

Start by searching online for poultry vets in your area. Organizations like the American Association of Avian Pathologists, the Association of Avian Veterinarians, and Poultry DVM have searchable directories by state. Search for avian, poultry, farm, or exotic specialties. Call local pet vets and ask if they treat backyard poultry. They may suggest others if they don't. Join local poultry groups online and ask for advice. Your 4-H leaders can also guide you. State agricultural extension offices often have poultry specialists who can consult, perform disease testing, and recommend vets. Even if a vet isn't experienced with chickens, a visit can help you learn and get medication if needed.

Knowing how to anticipate and spot trouble in the flock, build a resilient environment, and lay the essential foundation of care is key to better supporting our birds and, hopefully, preventing illness and injury. Prevention is only part of the puzzle. Prepare for the inevitable by building your first aid kit and having a veterinarian on hand in case you need help.

Feeding your flock healthy food is another piece of the puzzle in keeping them healthy. It is difficult for the immune system to respond or for the body to heal if it lacks the nutrients it needs to thrive. In the next chapter, I will discuss how you can ensure your flock receives adequate nutrition.

Lydia knows where to find the good stuff!

3

NUTRITION FOR A *Healthy Flock*

Nutrition is essential for a backyard flock. Their food and water are the building blocks of growth, egg laying, and immunity, including the ability to fight off disease. Just like with any species, nutrition is more than tossing them food. If your chickens aren't getting a balanced diet, you'll see dull plumage, egg-laying issues, weak birds, and even death. You'll also deal with vent gleet, bumblefoot, broken eggs, and a smelly coop. Chicken feed isn't one-size-fits-all, however. Nutritional needs change with life stages and circumstances. Even though their diets don't need to be perfect, it's important to have a basic understanding of their needs. Over the next few pages, we'll go over what your flock really needs and how you can best support their nutrition.

KEY NUTRIENTS FOR A HEALTHY FLOCK

EACH NUTRIENT IN A BALANCED chicken feed plays an important role. Humans wouldn't want to eat the same dry crumble every day, but chickens get excited to see their feed bag again and again. This makes it easy to provide the right nutrients with a consistent feed. Avoid making homemade feed, as it's hard to get the right nutrients from online recipes. Some ingredients need heat treatment, balancing, or pairing to improve absorption. Bulk ingredients degrade vitamins and fats over time, risking long-term issues. Chicken feed is well studied, with many options for affordability, organic, or sustainable needs, offering a balanced choice for every budget. Below you'll find some of the major nutrients your chickens need, what they do, and where they usually come from.

Protein and Amino Acids

Chickens need proteins made from amino acids to support muscles, organs, eggs, enzymes, antibodies, and feathers. Proteins must be in their feed to supply the correct amino acids. Some proteins are better sources than others. A protein deficiency can harm their health, causing poor feathering, fewer or no eggs, stunted growth, weak body, feather picking, a weak immune system, and low energy. The critical amino acids include methionine and lysine, often added as synthetic forms by feed producers to prevent excess protein.

I use starter feed when there are babies in the flock, but the chickens only care about how quickly I fill the feeder.

My Silkie and Bantams crowd around a feeder filled with their favorite treats.

Sources:

- **Soybean meal** is one of the most common plant-based protein sources in chicken feed because it is relatively balanced in amino acids.
- **Corn gluten meal** is a by-product of corn processing and is rich in protein. However, it is usually imbalanced and needs to be added appropriately.
- **Canola meal, cottonseed meal, rapeseed meal, and sunflower meal** are by-products of oil extraction, but they are often imbalanced and require supplementation.
- **Animal protein**, such as fish meal, meat, and bone meal, or dried insects can be used to supply protein and amino acids. The use of black soldier fly larvae is becoming more common in chicken feed. Fish meal is usually used sparingly to avoid fishy-tasting eggs.
- **Legume and pea plant protein** may be used if heat-treated to reduce antinutritional compounds.

Fats

Fats are an essential addition to chicken feed because they provide a concentrated source of energy, supply essential fatty acids, and support the absorption of vitamins. They play an important role in healthy cells and hormone production. Some fatty acids, such as linoleic acid, can't be produced by chickens, so they must be added to their diet. Fats also help limit feed dust and make the feed more appealing to the birds.

Sources:

- **Vegetable oils** can include flaxseed, soybean oil, canola oil, and sunflower oil.
- **Animal fats** can vary in quality but are used in some cases.
- **Insects and insect meal** are becoming increasingly common as sustainable sources of protein and natural fats.

Carbohydrates

Carbohydrates are the main energy source for chickens, fueling all activities. They require a steady supply to support digestion, immunity, and all body functions. This coop doesn't do low-carb diets! Note that fiber comes from carbs but must be balanced, as excess can hinder nutrient absorption.

Sources:

- **Grains** like corn, wheat, barley, millet, rye, oats, and sorghum are common cereal grains added to chicken feed.
- **By-products** such as brewers' grain, bakery waste, wheat bran, and grains left over at distilleries are often added to provide a more cost-effective carbohydrate source.

Sarah is one of my beautiful rescue hens who was dumped off at an automotive shop and was living under old cars until I was able to help.

Minerals

Both macrominerals, including calcium, magnesium, sulfur, sodium, potassium, chloride, and phosphorus, and trace minerals, such as zinc, copper, and selenium, are essential in a chicken's diet. Macrominerals support bones, muscles, cells, fluid balance, and nerves, while trace minerals aid in wound healing, feathering, blood cell production, thyroid function, and immunity.

Sources:

- **Calcium and phosphorus** are used for bone and eggshell support.
- **Mineral and trace mineral premixes** are used to help balance missing minerals in plant and grain ingredients.
- **Limestone or oyster shell** can be added for calcium, mainly for laying hens.
- **Sodium and chloride** are added for electrolyte balance, digestive function, and other things.
- **Inorganic trace minerals** are more commonly used but offer low bioavailability.
- **Organic trace minerals** are more bioavailable than inorganic forms, which means better growth and function.

My flock knows I am the source for feed. Notice the side eye as they wait for me to hand it over!

Vitamins

Vitamins are added in small amounts to support chicken health, including immune function, metabolism, growth, and reproduction. They are either fat-soluble or water-soluble. Fat-soluble vitamins include D (calcium absorption), A (vision and mucous membranes), E (immune support and antioxidants), and K (blood clotting). Water-soluble vitamins like B complex and choline aid digestion, nerve health, and metabolism.

Source:

- **Vitamin premixes** are added in various vitamin contents to feed to help balance what is missing from the already used ingredients.

FEEDING STYLES

Two of these 25-pound (11.3 kg) feeders will last all week, or more, for my Bantam flock!

You can spread feed in a shallow dish a few times a day for scheduled feeding.

THERE ARE A FEW OPTIONS for feeding your flock, and each has its benefits and drawbacks. Choosing one will depend on what works best for you and your flock. Consider space, flock size, personal scheduling, and free-ranging. Don't be afraid to change your approach after trying a method that doesn't work for you. Below are three common feeding styles, how they work, and the pros and cons of each.

Free-Choice Feeding

Leaving food out at all times for the chickens to take as needed is called free-choice feeding. Most people use large gravity feeders that you can fill and leave out. The feeders must be protected from the weather and critters.

Pros:

- Allows you to fill feeders to capacity and only fill as needed. Depending on your feeder size, this may not be very often.
- Reduces food dominance in the flock because they have all day to get to the feed.
- Birds can self-regulate and eat when hungry.
- You can leave for a vacation, and the chickens have food.
- No measuring or calculating feed.

Cons:

- Certain breeds could overeat (not likely with egg layers).
- Food waste and spillage depend on the type of feeder.
- Predators and scavengers could find food at night if not well protected.

This is the technique I use for my flocks because they do not have consistent access to pasture, and I keep many birds together. My run is very secure from mice, rats, and predators that may want to eat the feed. I use three gravity feeders that hold up to 50 pounds (22.6 kg) of feed and only fill them a few times a month.

Scheduled Feeding

In this method, food is given only at certain times of the day, such as morning and evening. The food is placed in small feeders, trough feeders, or pans, or tossed on the ground for the chickens to eat.

Pros:

- You have control over intake to prevent obesity in heavier breeds.
- There is consistent bonding and observation time with your flock.
- You save money with less food waste as long as you don't put too much down at once.
- If a sick bird is refusing food, you may notice sooner than with free-choice feeding.

Cons:

- Dominant birds may eat all the food before younger, weaker birds can get enough.
- Competition may result in pecking injuries.
- Chickens are dependent on your personal schedule. If you aren't available to feed them, they don't get to eat.
- Vacationing becomes more difficult because you have to find someone to feed them multiple times a day.

Foraging and Supplemental Feeding

This is when your chickens free-range and pasture-graze most of the day, and you give them feed to fill in the gaps. During free-ranging, they must have access to insects, greens, seeds, and grains. They will still need a balanced feed to supplement what they can't get from the land.

Pros:

- You save money on feed by utilizing the land.
- Chickens act as pest control for your property, as they will eat everything from spiders to ticks.
- Foraging is the most natural behavior for chickens and makes for a peaceful flock.
- Foraging provides entertainment and keeps them busy, reducing squabbles in the flock.
- Chickens help fertilize the land.
- It can be combined with free-choice feeding.

Cons:

- Soil fertility is needed for adequate nutrients.
- They need ample plant life and insects year-round. Best on large farms or pastures.
- Free-ranging increases the predator risk.
- Wild birds have constant access to the flock, increasing the risk of parasites and diseases.

SPOILAGE ISSUES Feed can spoil quickly and must be stored properly. Heat accelerates chemical reactions that break down nutrients. If the feed gets wet, it will become moldy and highly toxic to chickens. My favorite way to store feed is in an old, unplugged chest freezer. I found one secondhand, and even unplugged, it works well to keep my feed dry, cool, and away from mice. Steel trash cans with secure lids are great because mice can't chew through them. I suggest keeping them in a barn, shed, or garage so they are out of the sun and weather.

Chest freezer with feed

MONEY-SAVING TIPS FOR FEEDING A HEALTHY FLOCK

FEEDING IS THE HIGHEST COST in raising chickens (after you build that fancy coop). There are several ways to reduce feed bills and create a more self-supporting system. Thoughtful, sustainable strategies let you tailor your flock's eating habits to your budget and land. Below are a few methods you may adopt, along with the pitfalls you may encounter.

1. **Integrated Foraging**
   Creating a pasture offsets feed costs with minimal supplementation. Plant wheat, clover, kale, broccoli, legumes, and herbs for nutrients. Rotating chickens offers fresh forage but depends on seasons, weather, and space. Poor management risks predators and land damage. Over-rotating chickens can quickly clear land.

2. **Kitchen Scraps**
   Feeding chickens kitchen scraps is common. Food waste in landfills emits methane, harming the climate and water. Know which foods are safe for chickens, as noted at the chapter's end. Avoid overfeeding scraps; keep treats to 10 percent or less of their diet. Adult chickens eat about 4 ounces (120 g) of feed daily, so extras should be a few ounces per chicken. Despite not measuring, use this as a guideline. Too many extras can cause nutritional imbalance, lead to diarrhea, and attract predators.

Ricky forages in the yard and helps take care of the ticks.

My chickens love fridge cleanout day!

The magic door that leads to the garden

Sprouted seeds make the perfect treat.

Fermented feed is a wonderful gut boost.

3. **Chicken Garden**
   One of my favorite additions to my chicken area is a garden attached to their outdoor run. I block it off during the growing and harvesting seasons. After we harvest what we want, I open a chicken door to let the chickens in. By then, most of the garden is filled with weeds, mainly because of my busy schedule, but also because I know the chickens will love clearing it out. It takes my flock of thirty-eight about 5 months to clear the large garden, and by spring it's ready for planting again. They forage all fall and winter, and I get a cleared garden. Ideal plants include squash, garlic, herbs, sunflowers, berries, and brassicas.

4. **Fodder and Microgreens**
   Fodder is a hydroponic tray of sprouted grains that grows in about a week from seeds and water. Trays can be rotated for a daily supply, requiring only watering and cleaning. It can be scaled up or down. Microgreens are similar, fed earlier, rich in nutrients and antioxidants, and grown with or without soil, in trays or jars. No indoor space? Scatter seeds on the ground, water, cover with a kiddie pool, and in days you'll have greens for your chickens. Both are excellent when chickens lack access to greens.

5. **Fermentation**
   Fermenting feed is easy and cost-effective, requiring only water and a container. You can ferment any complete feed, but I prefer whole grain or mash to avoid mushiness. Rotate containers daily or do it as needed for sick or stressed chickens. Fermentation adds probiotics, boosts water absorption, and helps chickens feel full faster without causing nutrient deficiencies, as fermented feed is more bioavailable.

6. **Breeding Insects**
   Black soldier fly grubs have been taking over the chicken-keeping world for their high protein and calcium content! Chickens love to snack on dried grubs. People create miniature grub farms in sustainable systems, allowing their chickens to eat endless grubs!

7. **Bulk Feed Purchases**
   If you have a local farm or feed store that lets you buy feed in bulk at a discount, it could be your ticket to savings! Feed can spoil or lose nutritional value quickly, so pay attention to expiration dates to be sure you can use it in time. If not, consider splitting the order and the cost with a friend!

FEED PROFILES

Mash feed

Pellet feed

Crumbled feed

Scratch feed

NOW THAT YOU KNOW HOW you'll feed your flock, we can look at labels and what to buy. Choosing a feed form is important and, if I'm honest, a bit confusing at first. Feed forms (pellets, crumbles, and mash) affect waste, intake, and stability. Labels such as organic, medicated, and non-GMO can affect your chickens, the eggs, and even your wallet! To add to the confusion, feeds are formulated differently for different ages of chickens.

Feed Forms

Whole grain or mash: Whole-grain feed is the ingredients in chicken feed mixed together in whole-grain form. Mash is the same ingredients, slightly crushed. You can still see some of the grains or pieces of grain in the mash. Both forms can be a problem when the chickens pick out their favorite pieces and don't take in balanced nutrition. To cut down on this, you can mix the feed with water or ferment it before feeding it to the flock.

Pellets: This is feed that has been ground, pressed, and extruded into a consistent pellet shape. The pellets are designed so that chickens receive balanced nutrients with every bite. While pellets offer complete nutrition and minimal waste for adult birds, younger chickens may pick through and push out the larger pieces in search of smaller, crushed pieces. Pellets should be used only for adult birds.

Crumbles: Pellets that have been ground slightly or crushed into smaller pieces are called crumbles. Each piece provides the same balanced nutrition and is small enough for younger birds and chicks to eat. This can be fed to any chicken at any age.

Scratch: Scratch feed is not a complete, balanced chicken feed. It is a random mix of grains, corn, and seeds intended as a treat for the chickens. It is usually cheaper than balanced feed, but that is because it isn't formulated to keep your chickens healthy long term. Tossing it around in the yard or run is a great way to keep your flock entertained. They will scratch around and pick through the pieces. Because it isn't nutritionally balanced, it should be fed sparingly.

Feed by Age

Starter feed: This feed is designed to "start" your birds on the right foot. It is usually labeled for ages 0 to 8 weeks and is available in crumble or mash form.

Grower feed: This next step-up is for young birds that are 8 to 20 weeks old or until they start laying eggs. It is also available as a crumble or mash.

Starter/grower feed: Some companies combine the starter and grower into one feed that can be used from 0 to 20 weeks. It is available in crumble or mash form.

PRO TIP If you have a rooster in your flock, you can feed everyone an all-flock or grower feed instead of layer feed. Just make sure the hens have access to a feeder full of crushed oyster shells or limestone, and they will take them when their bodies need more calcium.

Layer feed: Once a hen matures and begins laying eggs, a layer feed can help her stay healthy. It contains slightly more calcium than starter and grower feeds to support eggshell development. Layer feed is best suited for laying hens, not roosters or younger birds, because the calcium can cause kidney damage, gout, and even death in young chicks.

All-flock feed: All-flock feed is formulated to support all chickens, whether they are hens, roosters, or chicks. It can also be fed to mixed flocks of different species. If you are integrating young chicks into the flock, all birds should be fed the safest food for the youngest members (all flock, starter, or grower).

Other Supplements

Grit: Chicken grit, made of insoluble rock such as granite and quartz, helps chickens grind their feed in the gizzard for proper digestion and to prevent blockages. Chicks fed crumble without outdoor access usually don't need grit unless they eat vegetation. Adults should always have grit available, though free-ranging chickens often find enough on their own.

Calcium source: Egg layers need calcium supplements, even if their feed provides calcium. They instinctively take what they need, preventing calcium depletion from their stores, which could weaken bones. I prefer crushed oyster shell, but limestone is also an option. Keep calcium separate from feed so hens only get it when needed.

Grit

Oyster shells

SPECIAL FEED LABELS

Organic: Organic feed is produced under guidelines and regulations that ensure it is free of pesticides, antibiotics, and fertilizers. This can be an excellent option for people who want to give organic eggs to their families or market their eggs as organic.

Non-GMO: These are feed ingredients that aren't genetically modified. Non-GMO doesn't mean organic, but the feed may be both if labeled as such.

Medicated: Some chick feed is available in medicated form. This feed contains amprolium, a medication that helps combat coccidiosis in young chickens. It can be used for a few weeks while they acclimate and gain strength. If they have been vaccinated for coccidiosis (no other vaccines matter), then they don't need medicated feed, and it can actually negate the vaccine. Medicated feed works best for chicks in their first few weeks of life, but I don't recommend long-term use. Prolonged use has been known to cause vitamin deficiency (primarily B1). You can find more information on coccidiosis in chapter 8.

SAFE EXTRAS AND TREATS

WE HAVE ESTABLISHED that a complete, nutritionally balanced feed is most important in your flock's diet, but what extras are safe or beneficial to feed them? I have seen people feed their chickens everything from raw deer carcasses to cake. Some things are fine in moderation, and others offer essential benefits. Here is a chart showing which treats, vegetables, or scraps can be given in moderation.

My chickens love it when I make them Chickuterie Tray . . . even though they would eat it if it weren't pretty.

FOOD ITEM	BENEFITS	DOWNFALLS	FREQUENCY
Leafy greens	Vitamins A and K, fiber, enrichment.	Greens spoil quickly. Too much can cause digestive upset. Avoid iceberg lettuce and stick to more nutrient-dense greens like kale, spinach, and cabbage.	A few times a week in small amounts.
Squash, zucchini, pumpkin	Vitamins, minerals, hydration, fiber, boredom busters. They can be stored long term in a cool, dry place.	Overfeeding can take away from nutrition balance. Many say pumpkin seeds will deworm chickens, but this is not true.	A few times a week. Great winter snack to avoid boredom.
Berries*	Antioxidants, vitamins, foraging favorites.	High in pesticides if not organic, expensive, high sugar content, possible choking hazard.	Occasional treat; feed as a small reward or scatter in the yard for foraging.
Melons	Vitamins, minerals, hydration.	High water content can cause loose stools; leftover scraps can attract pests or mold.	Feed on hot summer days to boost hydration or sparingly throughout the year as a boredom buster.
Apples, peaches, grapes	Vitamins, minerals; easily cut up and shared among the flock.	Apple seeds can be toxic in excess, peach pits must be removed, and grapes can be a choking hazard, so consider cutting.	Occasional treat.
Cooked rice, grain, or pasta	Offers fiber and carbohydrates.	Overfeeding can fill their body and reduce the amount of feed they eat. Never feed raw or dry to avoid swelling.	Occasional treat or kitchen scrap offerings.
Cooked eggs/ eggshells	Great source of protein and calcium.	I have yet to find a negative!	Once a week or more often if chickens need extra protein, such as during a molt.
Yogurt	May offer probiotics but not all studies show strong or even consistent evidence. Dairy isn't a natural food for chickens.	Dairy can cause digestive upset; yogurt must be plain with no added sugar or flavor.	Small, occasional spoonfuls.
Mealworms/ grubs	High in protein, vitamins, and fats. Food source chickens would naturally find in their environment.	Don't overfeed. Look for cleanly harvested sources without pesticides.	Can be given daily or a few times a week as long as it is less than 10 percent of daily diet.
Corn, scratch, mixed grain	Entertainment and encourages foraging behavior.	Can cause obesity and poor egg production. High in fat and carbohydrates.	Use sparingly.
Dried or fresh herbs	Vitamins, possibly antimicrobial effect but the proof is limited.	Could cause digestive upset for the amount chickens would need to have a medicinal effect. Essential oil would be more effective if the chickens are ill.	Small amounts.

*raspberries, blackberries, strawberries, elderberries, blueberries

UNSAFE FOODS AND TOXIC EXTRAS

SOME THINGS THAT SEEM HEALTHY and harmless can actually cause significant damage or even death. Chicken keepers must be aware of the dangers so they can avoid those items. Fortunately, chickens are good at avoiding toxic plants while free-ranging unless they don't have access to a food source or other options. Here is a chart showing toxic items, as well as things that should be fed with caution or in heavy moderation.

ITEM	RISK	NOTES
Moldy or spoiled foods	Mycotoxins are very toxic to birds, causing liver damage and even death.	Avoid all moldy feed and food scraps. Wet feed almost always turns moldy.
Avocado	Contains persin and causes possible heart damage, respiratory issues, and death.	The flesh of the fruit is fine, but most people avoid it for fear of accidentally including the skin and pit.
Apple seeds, fruit pits	Seeds and pits can release cyanide in digestion.	I have an apple tree by both of my chicken coops, and the chickens eat the fallen apples every year. We have seen no issues. Either it isn't enough to do damage or they are purposely avoiding the seeds.
Tomatoes, peppers, potatoes, other nightshade plants	Unripe fruit, leaves, stems, and roots can be toxic in excess.	Chickens will generally avoid as long as there are other options for food and forage.
Onions	Thiosulfate/sulfur compounds can cause anemia and hemolysis.	Small amounts may be okay but shouldn't be fed often.
Beans (raw/ undercooked)	Beans contain toxins that can be fatal for birds.	Cooked beans are okay to use.
Chocolate, caffeine, sweets	Theobromine, caffeine, and sugar can cause toxicity and death.	Avoid feeding at all.
Salty foods	Excess sodium can cause electrolyte and fluid imbalance and salt poisoning.	Don't make it a habit of feeding them salty foods.
Nonfood items*	Various toxic harm.	Avoid at all cost.

*rodenticides, lead, medications not prescribed, fertilizer, insecticides, copper sulfate, heavy metals, antifreeze, etc.

WATER: AS ESSENTIAL AS FOOD

Fresh, clean water is a must! Make it a habit to check their water daily and clean as needed.

These nipple attachments ensure clean water because chickens can't kick dirt into them. Genius!

WATER IS AN OVERLOOKED NUTRIENT, yet it is vital to every physiological process in your flock. Without adequate hydration, digestion and nutrient absorption suffer, egg production declines, and chickens lose weight, become lethargic, and get ill. A chicken cannot live without water, and it is our job to provide a constant, clean source . . . even though they will drink from muddy puddles!

How Much Water?

The average adult hen drinks about 2 cups (480 ml) of water each day. In hot weather, this amount can double or even triple. It's best to keep more water available than needed to prevent running out. I also suggest keeping several waterers on hand in case one falls over or leaks. I'm not sure what my chickens do in that run when I'm not around, but they have managed to tip over a 5-gallon (19 L) bucket of water! I always keep extra so they can still drink until I notice the spill. Most waterers leak if not kept level, and chickens are good at bumping them just right so all the water slowly drips into the ground.

Here is a basic calculation to estimate how much water your flock will need. Keep in mind that your flock may drink more or less, and this is just an estimate.

_____ number of birds ÷ 8 = _____ gallons per day
_____ gallons per day x 2 = ______ gallons on a hot day

I always keep 3 or more days' worth of water available. This ensures the chickens still have water when I am forgetful or when my schedule prevents me from doing my usual daily chores. For my flock of thirty-eight chickens, I keep four waterers, each 5 gallons (19 L).

38 ÷ 8 = 4.75 (I round up to 5 gallons [19 L] per day.)

On hot days, they will need up to 10 gallons (38 L) a day. I also keep extra as backup in case of spills or if I am late to refill. My chickens rarely go through this much water, but it is always better to have more than to run out.

Here are a few basic water tips:

- Keep waterers raised off the ground so they stay cleaner.
- Place waterers in the shade to prevent algae growth and keep water cool.
- Secure waterers away from wild animals, such as wild birds, to prevent the spread of disease.
- Check the water daily, even if you usually don't refill daily.

Of course, no matter how clean you keep their water, they will find the dirty puddles.

Water Management

It seems straightforward: Fill the water, let them drink it, and fill it again. But clean water isn't just about how it looks. Waterers are among the most likely places for bacteria to grow and for germs to spread through your flock. Water can freeze quickly in winter, leaving your chickens dehydrated. Chickens kick around bedding and substrate, creating very dirty water.

Here are my best water management tips:

- Use enclosed waterers with nipple drinkers to prevent debris from entering the water. In the summer, I switch to cups so they can get more at a time and dip their beaks in, but these need to be cleaned often.
- Rinse the waterer of debris during each refill, and clean it with soap and water weekly or as needed. I find that a toilet brush, used only for chicken waterers, works well for scrubbing.
- Have a plan to prevent water from freezing in winter with submersible heaters, heated waterers, or wraparound heaters. These will need to be plugged in, and if you don't have electricity in your coop, you'll need to run an outdoor extension cord to the coop. If this is not an option, plan to visit the chickens multiple times a day to break the ice.
- In hot weather, keep waterers cool by placing them in the shade, adding ice, or refilling often.

Water Additives

On a normal day, I give my chickens fresh, clean water. However, there are times when we need to add supplements or medication to the water. Well-balanced feeds are meant for everyday use, and special circumstances may require supplementation. Let's look at common water supplements and when they may be needed.

- **Enzymes:** Enzymes are added to feed to help break down compounds that may be difficult for a chicken to digest. Adding enzymes improves digestibility, reduces feed waste, and lessens the need for many mineral supplements. Many water additives include enzymes. Digestive enzymes can be used during periods of stress or change, during illness, or during those first few fragile weeks of life.
- **Prebiotics and probiotics:** Probiotics are the "good bacteria," or beneficial microorganisms that support a healthy gut by reducing harmful bacteria. They also support immune health. Prebiotics are compounds that feed those beneficial microorganisms. These supplements can be beneficial during periods of stress or illness and after antibiotic use.
- **Electrolytes:** Electrolytes are minerals that carry an electrical charge when dissolved in the body. All animals need them for a range of bodily functions. They help with fluid balance, hydration, pH balance, metabolism, enzyme function, heat stress, recovery, and nutrient absorption. Overusing electrolytes or using them for an extended period can cause many issues, including kidney stress, salt toxicity, and other imbalances. It is best to use them only as needed during times of stress, heat exhaustion, recovery, diarrhea, or transportation.

- **Vitamins and minerals:** Vitamin and mineral mixes should be viewed as a short-term boost rather than a daily need, especially when providing a balanced feed. They can be given during times of stress, such as travel, disease, and environmental changes. They can also be helpful when chickens are eating less because of heat or illness. Some medications can deplete the system of vitamins, so adding them back to their water for a few days is essential.
- **Medication:** Antibiotics, dewormers, and other medications are sometimes added to the water source. Follow the instructions on the packaging for best results.

Feeding your chickens a balanced diet can seem overwhelming, but hopefully you now have a clearer understanding of the choices and process. But no matter how much effort you put into feeding your flock the right foods, a predator can undo all that hard work in a single night. In the next chapter, we'll shift from what goes into our chickens to the dangers that lurk around them. I'll give you solid options for keeping your flock safe with true predator protection. I'll also share common predator profiles so you can narrow down which type of animal may be bothering your flock.

My flock is enjoying some outside time, eating bugs and grass, and knocking over a few flowerpots.

Beatrice exploring the yard knowing the rooster is nearby to warn her about dangers

4

PROTECTING YOUR *Flock*

The chicken my son first bonded with, Nugs, died while we were away. We were on vacation, and her life was snatched up, literally and figuratively, by a hawk. My son, then four, was devastated. He is eleven now and still talks about Nugs. She used to ride in the back of his tricycle, and she was the only one who let him carry her around. Unfortunately, our run at the time wasn't secure. I trusted that "nothing bad has happened in a while" and that it probably wouldn't while we were gone.

Chickens aren't just poultry or livestock . . . not to us, anyway. When we bring an animal into our life, we care for it like a pet. Chicken keeping often starts as a way to get fresh eggs and turns into a hobby that brings joy and comfort. However, you don't have to be attached to your flock to want to keep them safe. Defending your budget, livelihood, or food source is just as important.

Protecting your flock goes beyond a secure, predator-proof coop. A balanced diet and all the love in the world won't protect your flock from raccoons that can pry open chicken wire, winter ice storms, or parasites that go unnoticed. My goal is to map out the threats and help you establish solid husbandry, because you can't protect against what you aren't prepared for.

My son, Henry, with Nugs as a young chick

PREDATORS

THERE ARE TWO SCHOOLS of thought in the chicken-keeping world when it comes to predators. One leans toward eliminating predators by trapping or shooting them. Unfortunately, by the time you know a predator is lurking around, it is too late to kill it because it has already killed or harmed members of your flock. It is a constant battle of damage control after something tragic has already happened. Another issue with this method is that even if you dispatch one predator that took your hens, another is waiting nearby to take more.

A more productive method is to secure your flock at night and take precautions during the day that keep predators from getting your birds in the first place. I live on 5 acres (2 ha) next to about 100 acres (40 ha) of wooded land. I know there are predators like raccoons, skunks, coyotes, and minks lurking around my coop at night, but I am confident in our coop design, so I know my chickens are safe. During the day, the biggest predator we face is hawks. While we take care to set our flock up for success, hawks occasionally harm our birds, but over the years, we have been able to greatly reduce this.

Each region has a different set of predators that attack or bother a chicken flock. It's important to know what threats are present in your area, so I encourage you to join local chicken-keeping groups, ask members, or search files to become familiar with local predators. Your best bet for keeping your flock safe is to know what you're keeping them safe from. One important thing to note is that each state or county has laws and regulations about how you handle predatory animals. It may not be just a question of ethics but also a question of legality. Before you do something hasty, like shoot a hawk from the sky, check with local officials to make sure you aren't breaking any laws.

After a run-in with a predator, Maple is still thriving with me today, and she might just be the bossiest of the whole group!

A Safe Coop

My biggest regret about letting my guard down is the story of Maple. Maple, a Mille Fleur D'Uccle Bantam, is friendly and full of personality. As she and her brooder mates outgrew the garage, I wanted to give them more space. The coop I was building wasn't ready, so I bought a cheap big-box-store coop. I set it up in the yard, brought the pullets (young hens) out in the mornings, and returned them at night, giving them sunlight and forage. I knew it wasn't safe overnight, so they stayed in the garage. One evening, we returned late. The coop's bottom tray was pushed out by a raccoon that had grabbed Maple's leg, severing it. I thought she was a goner, but chickens heal well, and she's now bossing her flock. Many coops on display aren't safe—they have flaws that can lead to tragedy. Use this checklist when building or buying to ensure safety.

I hate to remove animals from their homes, so if I can deter them instead, I will. If they must be trapped, relocation is the best option if it is legal in your area. Avoid relocation during baby season so you don't take a mother from her babies.

Hardware cloth needs to be secured with washers and screws. This keeps predators from pushing out the staples.

- Put your coop in a safe location. It should be in direct view and away from predator-dense areas, if possible.
- Coops should be built from strong wood with sturdy walls, studs, rafters, and strong plywood. The cheap wood used in put-together coops won't withstand a neighbor's dog or a predator's chewing.
- Windows should be covered with hardware cloth (also called rodent mesh) secured with screws and washers, not screens. That way, when the windows are open for airflow, predators can't get in.
- Elevate your coop off the ground so the bottom doesn't rot or weaken. I use a layer of gravel, 6x8 treated boards on top of that in a square shape, hardware cloth covering the middle, and set the coop on top. That way, critters aren't able to live under the coop, and it is away from the moisture of the ground.
- Use latches and locks that a child can't open to ensure a raccoon is also unable to open it. I use two latches on my coop door so there is always a backup.
- Add an automatic door so it closes even if you aren't home to put them away. Most doors open and close with the sun. There are even doors you can open and close from an app.
- Make sure the ventilation areas are covered with hardware cloth. If a rat or mink can get it, they can harm your chickens while they sleep at night and have nowhere to go.

A Safe Run

Daytime chicken protection is trickier than at night. Locking them up at night is easy, but during the day it feels harsh. We can provide safety options and set them up for success, but predators can still breach defenses. Our attached run has an automatic door opener, letting the chickens go out, protected yet free to move. The run should be large enough, about 10 to 15 square feet (1 to 1.4 square m) per chicken, so all are happy. Here are safety tips for your run.

My roof is made from scrap materials my dad collected from his business.

- Build your run with footers that are made from treated wood and cemented into the ground. This protects them from rotting and prevents anything from lifting it. Use treated wood around the base and on the walls.
- Cover the entire run with hardware cloth. Chicken wire keeps chickens contained, but predators can still reach through or break through. Use screws and washers to secure the hardware cloth to the wooden crossbeams, and be sure to overlap it.
- Your chickens will need an overhead covering. I strongly suggest a roof because it also keeps them safe from the weather. Hardware cloth is the next-best option, but bird netting can also be used. Bird netting will need to be replaced more often and is still weak enough that something could get in (especially nighttime critters looking to eat your chicken feed).
- Skirt out the hardware cloth along the base of the coop a few feet and cover it with soil and landscaping. This prevents diggers from getting under the run.
- Secure run doors with double latches or locks.
- Add a solar-powered camera with motion detection so you can be alerted to nighttime entries or at least review footage to see what may have gotten in.
- Remove hiding places around the outside of the coop and check the perimeter often.
- Secure feed or store it somewhere away from the coop and run. Remove food scraps before nightfall to avoid attracting animals.

FENCED AREA This run is protected by heavy tree cover. If you don't have trees, you could use bird netting to keep hawks and other aerial predators at bay. Most chickens will stay within a fence that is around 5 feet (1.5 m) tall. My fence is 4 feet (1.2 m) tall, but I have a rooster who prefers to stay with them, and he can't jump the fence. If I have a hen who won't stay in the fence, I will trim her flight feathers on one wing to keep her grounded. (This is much like trimming a human fingernail if done properly.)

This chicken finds safety in the trees and brush.

My goats and geese are housed apart from my flock, but I often let them out during free-range times.

This owl decoy helps keep hawks and other birds away.

- Even with hardware cloth, predators in some parts of the world can still reach the flock. If you live with larger predators such as bears or coyotes, consider using electric fencing to protect your flock.
- If you don't free-range your flock, they will appreciate some safer time outdoors. My large coop has a chicken run attached, as well as an outdoor run. This run doesn't have a structural covering, but it does have many trees and underbrush to help keep them safe. You could also use bird netting.

Safe Free-Ranging

The title "Safe Free-Ranging" is a bit overconfident. We can't keep chickens 100 percent safe while they are free-ranging, but I think it is a necessary compromise to allow them to live naturally. Nothing makes a chicken flock happier than having complete access to the outdoors, where they can find insects, dig holes, dust bathe, and eat vegetation. Every chicken keeper who lets their flock out to free-range knows it is dangerous.

Here are a few steps you can take to give them the best chance of avoiding and outsmarting predators.

- The number one thing your flock needs when free-ranging is a rooster. A good rooster will watch the area and the sky for predators while the hens forage. It is a beautiful sight to see the hens, heads down, oblivious to their surroundings, while the rooster has his head up and is paying attention. He will call to the hens if he sees anything, and they will run to cover. If something attacks, a good rooster will risk his life to neutralize the threat.
- Give them access to free-range space with ample tree coverage. Hawks and the like usually won't swoop down under trees because it isn't a quick getaway, so chickens tend to stick near the tree line in case an overhead predator is spotted.
- Provide structures for them to run under, such as picnic tables, shade cloth, swing set structures, or bushes.
- Enlist the help of livestock guardian dogs or guard geese. My goats are defenseless against predators, but even the sight of them keeps predators at bay.
- Set up an owl, dog, or coyote decoy.
- Befriend wild crows, who will chase away hawks. It is a pretty interesting sight to watch a crow go up against such a major predator!
- Hang shiny items in trees around their area. String CDs onto wire or rope and run it horizontally across their space.
- Work on recall training so you can quickly call them back to safety. A few days of shaking a bag of dried grubs while you make a specific call will show them how it's done.
- Use electric solar fencing around a large area to keep out daytime predators, such as foxes. You can move the fencing regularly to provide fresh foraging space.
- You may choose to let them out only when you are available to supervise. Most predators won't encroach on the area if you are there.

Common Predator Profiles

BEAR

Danger Time:
Dusk, night, early morning

Style of Attack: Brute force. Bears smash, tear, or claw their way in. Not quiet.

Likely Injuries:

- Entire bird carried off or eaten on-site
- Partially eaten carcasses or crushed remains

Entry Methods:

- Unsecured doors and windows
- Weak fencing
- Weak coop walls
- Thin siding or weak roof

Signs of Presence:

- Major damage to structure (wire ripped, doors broken, walls smashed)
- Feathers and debris scattered
- Large claw marks, deep gouges on structures
- Large paw prints in the dirt
- Scattered garbage, knocked-over bins

Prevention Tips:

- Build coop with heavy, reinforced material.
- Use electric fencing around the perimeter of your yard or coop.
- Secure trash, compost, and feed, especially at night.
- Use secure gates and locks.

DOMESTIC DOG

Danger Time: Anytime

Style of Attack:
Rushing in quickly, bites, mauling, chasing

Likely Injuries:

- Many birds maimed or carried off
- Blood and feather trails

Entry Methods:

- Weak or unlocked gates
- Loose fencing or hardware cloth that's not secured
- Gaps under fencing
- Low fencing the dog can jump over

Signs of Presence:

- Broken mesh at an entry point
- Dog tracks
- Bent or damaged fencing
- Feathers or carcasses near entrance or scattered

Prevention Tips:

- Secure gates and fencing.
- Use taller fencing without gaps.
- Add motion-activated alarms.
- Keep guardian animals.

OPOSSUM/SKUNK

Danger Time: Night

Style of Attack: Opportunistic scavenger. Will take eggs, young chicks, or weak birds. Rarely attacks adults.

Likely Injuries:

- Missing eggs or eggshells with bite marks
- Missing or partially consumed chicks
- Internal organs, shells, or wet feathers left behind

Entry Methods:

- Gaps under run floor or in coop structure
- Screens or mesh that is easily broken
- Open doors
- Rotting wood

Signs of Presence:

- Eaten chicken feed
- Eggshells
- Skunk or opossum paw prints
- Skunk odor or droppings
- Disturbed bedding or nesting material

Prevention Tips:

- Seal gaps, openings, and foundation.
- Block any burrowing spaces.
- Lock doors and external nest box entry.
- Eliminate food scraps.

RAT/RODENT

Danger Time: Dusk, dawn, mostly night

Style of Attack: Gnaws through weak structures. Usually an indirect predator who is after eggs, scraps, and sometimes chicks.

Likely Injuries:

- Broken eggshells
- Missing chicks or partially eaten chicks

Entry Methods:

- Holes and gaps (even smaller than 1 inch [2.5 cm])
- Cracks in the structure or floorboards
- Gnawed chicken wire

Signs of Presence:

- Gnawed wire, wood, mesh
- Teeth marks anywhere
- Noises and rustling at night
- Missing feed
- Tunnels
- Nests in bedding
- Large droppings

Prevention Tips:

- Use ¼-inch (6 mm) hardware cloth.
- Seal all cracks, gaps, and openings.
- Store feed in rodent-proof containers.
- Remove feed and scraps before nightfall.
- Don't use poison, as the chickens could eat it.

DOMESTIC CAT

Danger Time: Dawn, dusk, night, daytime

Style of Attack: Stalk and pounce small chicks or small Bantam breeds

Likely Injuries:

- Small chicks missing, pieces of flesh missing, puncture wounds
- Partial remains if it is too big to consume or carry away

Entry Methods:

- Large gaps in walls or mesh
- Open doors or gates
- Climbable posts or fencing

Signs of Presence:

- Cat prints
- Feathers stuck in wire
- Patrial carcasses, carcasses that have been regurgitated, or scattered body parts
- Cat fur stuck in fencing

Prevention Tips:

- Secure sides and top of structures with mesh or netting.
- Remove climbing structures or secure the top of areas.
- Lock up birds securely at night.
- Don't allow them to free-range with cats around, especially if there are chicks in the flock.

Note: My barn cats were raised around my chickens and do not bother them. They will lay next to them while they free-range in the yard. I am more concerned with strays or neighbor cats.

BOBCAT

Danger Time: Dusk, night

Style of Attack: Stealthy ambush, stalking, leap/pounce

Likely Injuries:

- Missing chickens, but not always feathers left
- Minimal disturbance (carry away prey to a safe location)
- Puncture wounds and claw marks on the remaining chickens

Entry Methods:

- Gaps in fence, structure, and roof
- Birds fenced in without overhead coverage, allowing bobcats to climb into the fence
- Low fencing
- Weak roof or tops of runs

Signs of Presence:

- Large cat prints
- Missing birds with no evidence of struggle
- Bobcat scat or territorial markings

Prevention Tips:

- Build structures with tall fencing and predator-proof tops.
- Remove nearby climbing structures.
- Reinforce all gates.
- Use electric fencing around the perimeter.

RACCOON

Danger Time: Nighttime, Dusk
Style of Attack: Clever, uses paws like hands to unlatch doors or pull apart fencing

Likely Injuries:

- Missing heads or body parts
- Severed limbs or wings
- Scattered feathers
- Crop or chest torn away

Entry Methods:

- Loose or flexible wire pulled apart from the structure
- Doors or windows that can be opened
- Latches that a child can manipulate
- Uncovered vents

Signs of Presence:

- Muddy paw prints on windows and the structure
- Open latches that were secured
- Dirty water sources, as raccoons like to wash their paws
- Claw marks
- Feathers suck in openings where chickens were pulled through

Prevention Tips:

- Use hardware cloth rather than chicken wire.
- Bury wire around the edge of the coop and run.
- Block all openings a small paw could fit through.
- Use motion lights or alarms.
- Use indoor nesting boxes only.
- Remove feed before dark.

MUSTELIDS (MINK, WEASEL)

Danger Time: Mostly nighttime
Style of Attack: Stealthy, persistent, slips in spaces without notice of destruction

Likely Injuries:

- Birds appear to be killed as sport, with multiple bodies left behind
- They make small wounds and drink the blood or eat the organs
- Vent damage or entrails removed

Entry Methods:

- Gaps or tiny spaces anywhere in the coop
- Gaps in corners or closures
- External nest box that doesn't seal tightly

Signs of Presence:

- Multiple dead chickens in one night
- Blood trails or skid marks in bedding

Prevention Tips:

- Build with hardware cloth, not chicken wire.
- Seal all gaps, cracks, and corners.
- Elevate the coop and secure the base.
- Limit the rodent population with barn cats because weasels often follow rodent trails.

COYOTE

Danger Time: Anytime, mostly at night
Style of Attack: Similar to domestic dogs; they ambush, dig, and force entry through weak spots

Likely Injuries:

- Severe tissue or muscle damage
- Neck puncture wounds
- Internal bleeding
- Broken bones

Entry Methods:

- Gaps or weak structure in gates or fencing
- Dig under fencing

Signs of Presence:

- Blood at kill areas and entry points
- Damaged fencing, gates, and wire
- Dig spots around the fencing
- Feathers leading away from the coop
- Howling at night
- Paw prints
- Disturbed soil where coyotes have dug

Prevention Tips:

- Bury hardware cloth in an apron around the base of the coop and run.
- Keep structures strong and intact.
- Secure gates with strong latches.
- Clear brush where they can sleep and hide.

HAWK

Danger Time: Daytime

Style of Attack: Swoops down from above, grabs the bird with talons. Doesn't fly around in circles like buzzards, but rather swoops back and forth or waits in trees for an opportunity.

Likely Injuries:

- Missing birds with a few feathers left behind in one spot
- If scared off, birds may be left behind with talon holes or gashes
- The head and neck are sometimes removed

Entry Methods:

- In the yard during free-ranging
- Open runs or fencing

Signs of Presence:

- Screeching sounds while soaring
- Visibly spotted in trees or flying overhead
- Feathers left in one spot and a missing chicken

Prevention Tips:

- Cover runs or open fencing with bird netting or a roof.
- Provide hiding places for chickens (picnic tables, bushes, shade cloths).
- Use reflective deterrents or streamers.
- Put an owl decoy on top of your coop or run.
- Use roosters as protection.

OWL

Danger Time: Late evening until dawn

Style of Attack: Attacks silently from above. Not as common because chickens are usually inside at night.

Likely Injuries:

- Missing heads
- Severed necks
- Puncture wounds from talons

Entry Methods:

- Open runs, gaps in roofing
- Chickens accidentally left out at night or sleeping in trees
- Uncovered sleeping space

Signs of Presence:

- Missing chickens
- Owls are heard at night
- Feathers/blood left at the site

Prevention Tips:

- Cover runs with wire, roofing material, or strong bird netting.
- Lock chickens inside the coop at night.
- Check cameras inside the coop or do a head count after dark.

EAGLE

Danger Time: Daytime

Style of Attack: Circling the skies above before swooping down and grabbing prey. Chicken is usually carried off.

Likely Injuries:

- Birds are gone, with possibly a few feathers in one spot
- If found, they have deep talon wounds and torn flesh

Entry Methods:

- In the yard during free-ranging
- Open runs or fencing
- Areas with little coverage and trees nearby

Signs of Presence:

- Eagle feathers found
- Large talon wounds
- Soaring eagle near the area (lock up flock if spotted)

Prevention Tips:

- Let chickens forage in a covered area.
- Use reflective material, moving deterrents.
- Strong and well-attached bird netting.
- Roosters or guard dogs/geese may be helpful.

SNAKE

Danger Time: Dusk until dawn or during warm daylight hours
Style of Attack: Hide in coops, swallow eggs or chicks

Likely Injuries:

- Missing eggs, hens refusing to use the nest box
- Missing whole or partial chicks
- Adult birds are usually unharmed

Entry Methods:

- Cracks or gaps in the coop
- Coops left open during the day
- Snakes can fit through small holes and quickly hide in bedding or nests

Signs of Presence:

- Holes in shells or missing eggs
- Snake skin near or in the coop
- Slide marks left in dirt, dust, or bedding

Prevention Tips:

- Seal gaps top to bottom.
- Cover openings with hardware cloth.
- Secure the underside of the coop and check for rotting wood.
- Keep the area around the coop free from brush and debris (snakes love an old pile of hay).

PERSISTENT PREDATORS Some predators are very persistent. They will return again and again to find the weakness in the flock or in the setup. The best thing you can do is lock your chickens in the coop and the run for a week or two until the predator moves on. You can also change your defenses daily to confuse them or set up trail cameras to catch them in the act. Once you know who they are and what they are trying to do, you can adjust accordingly.

My chickens forget about everything when snacks are involved!

Supporting the Flock After an Attack

Chickens are resilient, and for the most part, you won't notice a lot of distress after an attack. If you aren't there to see the attack, you may not even know which ones were around when it happened. There are a few steps to restore security and health to your flock.

1. Remove the remains and sanitize to prevent stress and not attract predators. Use a poultry-safe disinfectant to clean the area.
2. Most predators will return once they have made a meal of your chickens and know the entry points or routines. Modify weak areas.
3. Keep your chickens confined to the coop and run for a few days or a few weeks to discourage the predator and encourage it to move on.
4. Look for burrows, nests, and hiding spots where they could be lurking. Cover free-ranging space with bird netting or provide large structures for hiding. Add cameras to spot the predator.
5. A stressed flock may be more skittish or refuse to leave their covered run. Offer treats and enrichment as a distraction. Consider a few days of electrolytes in their water and expect a few days of slow egg-laying.

WEATHER PROTECTION

I always wrap my run in plastic to block the wind so the birds can use their feathers to hold in heat. I use plastic roofing panels or painters' plastic.

PROTECTING YOUR CHICKENS from the weather can be easy with the right setup, but if you don't account for all the seasons when building their coop, you may struggle when the weather changes. For years, I spent time adjusting my coop because things didn't work or hold up in the weather. Learn from my mistakes with my best weather tips!

Cold Weather Tips

Winter is the season that tends to make most new chicken keepers anxious. Maybe it's because we don't enjoy staying outside in the cold for long periods. Keeping chickens warm isn't your job. Your job is to keep them dry and out of the wind. The chickens will handle the rest. Here are the only things I do to keep my flock happy in the winter.

WIND PROTECTION

Chickens keep warm if they have a dry, wind-protected shelter, day and night. During the day, a covered run with plastic sides helps keep them dry and shielded from the wind. While they can go outside, they prefer the dry run in cold or snowy weather, or they'll stay inside if no run is available. Staying indoors can lead to a buildup of droppings, health issues, moisture, spills, and pests. Even if chickens free-range in warm months, having a run or barn for winter is advisable.

My chickens really appreciate this extra warmth when it is below freezing, but Bluey is hoping it will dry out her wet hair after getting caught in the rain!

HEAT

Most climates don't need heat for chickens, and it can even harm them. Chickens regulate their temperature well, and heat sources can confuse them. Warm nights with cold days can shock them. Their down feathers trap heat, and a draft-free coop suffices. In Ohio, with temperatures down to -15°F (-26°C), I kept chickens outside without heat, but now I use overhead or side radiant heat to prevent frostbite, and it also slightly warms the air. In colder areas, extra help might be needed. Chickens often tuck their heads under their wings to keep warm. I advise against heat lamps due to fire risk, and they are often unnecessary.

INSULATION

While I have had coops in the past that were not insulated, I find that insulating them is beneficial. A few years ago, I got a large shed and converted it into a coop. I used thick foam board insulation between the studs and covered it with plywood. This coop stays about ten degrees warmer than my old one. It is also beneficial in the summer months, as it holds the temperature steady instead of getting too hot.

FOOD

If you feed your chickens a few times a day during the warm months, consider putting out free-choice gravity feeders for the winter. They will need a steady supply of food to stay warm. I give my chickens a higher-protein feed during the fall and winter to support feather regrowth and help them stay warm. You can also add extra carbohydrates, such as corn, as a bedtime treat on very cold days.

WATER

Your flock will need a water source all day long. This can be a problem when temperatures are below freezing and their water turns to ice. I use a drop-in or wraparound heater to keep the water from freezing. My coop does not have electricity, so I have to run an outdoor extension cord to my run. If this is not an option, you will need to check their water several times a day to make sure it has not frozen over.

Hot Weather Tips

In my opinion, the most concerning weather for chickens is heat. It is much easier for chickens to stay warm than to cool down. They don't have sweat glands like humans. Chickens cool themselves by panting, spreading their wings, or seeking shade. Overheating can cause heat stress, which can be very detrimental. There are a few ways you can make sure your flock stays cool.

SHADE

This is the most important thing you can provide for them. If you are starting out, put your coop at least partially in the shade. Trust me on this one! Not only will shade keep chickens cool enough to handle the heat, but it also prevents sunburn on their combs and wattles. A covered run and big trees are great ways to provide shade for your chickens, but if you don't have those, consider adding shade cloths or other structures.

AIR CIRCULATION

If you notice a lot of panting or pale combs, it might be time to help your chickens manage the heat. Offering a few large fans in their outdoor areas and in the coop can help cool the area. You can also add a ventilation fan to the top of the coop to help pull out hot air. Of course, having windows in the coop that you can open day and night is a great addition.

WATER

Your feathered friends need access to fresh water all day long. They rely on water to stay hydrated and regulate their internal temperature. Without water, their organs become stressed, their appetite decreases, egg production is suppressed, and in severe cases, they can suffer from heat exhaustion or even die. I suggest keeping water out of the sun to help it stay cooler and prevent algae growth. Fill their water daily to keep it fresh and clean.

Lucky's open-mouth breathing and wings held out are telling me she thinks it's too hot.

Caswell got caught in the rain! She doesn't seem to mind.

Even though chickens love the mud, it isn't great for their health to live in it.

Wet Weather Tips

Storms, rainy seasons, flooding, and high winds can really put a damper (get it . . . damp-er?) on chicken keeping. There is nothing worse than wet chicken poop and bedding. Not only is it gross and smelly, but it can also cause injury, illness, and bacterial growth. The best thing you can do to avoid a damp coop and run is to make sure it is built with a roof, but there is more to it than that.

OVERHEAD COVERAGE

Start with a roof on your run so most of the rain stays out. As a bonus, it will also keep debris out during storms. I love having my coop in the shade to keep it cool in the summer, but that also means old branches and twigs could fall.

EXTRA WATER PROTECTION

It's important to keep running water out of your run. The roof will protect it from rain above, but if your yard is known to hold water or your coop is at the bottom of the hill, water will rush in through the sides. Our house sits on a hill, and the rest of our land is at the bottom, so our coop is at the bottom. We installed French drains to reroute water away from the coop during heavy rain. We also installed gutters on the run. This helps keep rain from blowing in through the sides of the run.

WIND PROTECTION

I have heard of box-store coops being blown away in the wind. It is essential to secure coops to the ground before a big storm harms your chickens, who might be trapped inside. Even if a coop is too heavy for you to lift, the wind can still be powerful. You can use ground anchors if you don't have posts cemented into the ground.

DISEASE AND PARASITE CONTROL

ALL CHICKENS KEPT OUTDOORS are at risk of disease and parasites. There are practical steps you can take to prevent these problems, but you can't avoid them altogether. If you plan to raise chickens, be prepared to handle common ailments such as viral infections, mites, lice, and worms. Chicken keeping is more enjoyable when these problems occur less often or are short-lived.

Biosecurity

Biosecurity involves measures to prevent the entry and spread of pathogens and parasites, such as barriers and protocols that protect your flock. Outbreaks can cause financial and emotional harm. It helps prevent diseases such as avian influenza (page 119), Newcastle (page 117), respiratory viruses (page 113), and Marek's (page 116), which can originate from wild birds, people, contaminated materials, other farms, and equipment. Here are key practices to remember.

- Limit and control visitors to your flock. Humans can carry pathogens from one farm to another on tools, clothing, shoes, and vehicles. Visitors with a high likelihood of exposure to other flocks should not wear the same boots and clothing around your flock. Have them use a boot wash before entering your flock's area.
- When visiting places with potential exposure, you should shower and change your clothes. Don't wear your chicken boots to another home or farm with chickens. If unavoidable, use a boot wash upon return.
- Quarantine all new flock members for up to 4 weeks. Observe for signs of illness or parasites before introducing them to your established flock. Existing flock members who get sick should also be quarantined away from the flock.
- If you have multiple flocks, you can limit contamination risks by cleaning tools, feeders, waterers, and boots before moving from one flock to another. I usually don't do this unless I know one of my flocks is sick.
- Keep rodents, pests, and wild birds out of your chicken enclosures by using hardware cloth instead of chicken wire, sealing feed storage, and reducing hiding places around the coop.
- Limit exposure to wild waterfowl by keeping your coop away from large bodies of water or out of the path of flyover zones if there are lakes and ponds nearby.
- Conduct daily inspections of your flock to detect problems early. Keep detailed records of vaccinations, medications, and illnesses within your flock.
- If you have a flock affected by disease, do a complete cleanout and disinfection of the coop and run. Let it rest and dry for 90 days (6 months is preferred) before introducing new birds.

BOOT WASH DIRECTIONS Use a bin you can easily step into and place a plastic scrub mat at the bottom. Fill the bin with a disinfecting solution to a depth of about 1 inch (2.5 cm). The solution can be a 1:4 bleach-to-water mixture or a poultry-approved disinfectant. Remove debris from the boots before disinfecting. Step into the bin to disinfect. Keep the boots in the solution for 1 to 5 minutes, following the directions. Rinse with water if you are going directly into the chicken space or to protect the boots from damage.

Matilda and her friends stay safe with thoughtful biosecurity measures.

ASK THE POULTRY DOC
PARASITE INFECTION

Chickens do commonly carry low-level parasite burdens, and the goal is not to make birds "worm-free," but to prevent clinical disease. Indiscriminate, frequent deworming is inappropriate and can contribute to resistance if done incorrectly.

Conversely, relying on visual confirmation of worms in droppings is a dangerous strategy. Many clinically significant parasites, such as Capillaria and cecal worms, are microscopic or rarely visible to the naked eye. By the time you see adult worms, the infestation is already severe, often causing weight loss, diarrhea, poor condition, or catastrophic disease. The majority of the parasite burden consists of larval stages that migrate through tissues, causing inflammation and damage well before adult worms ever appear.

When diagnostics aren't available, a targeted, limited course using a labeled product at the correct dose is a reasonable, evidence-based alternative. This is not routine or preventive monthly worming; it's a risk-based intervention.

Unlike unproven remedies, approved anthelmintics are backed by rigorous clinical trials that establish their efficacy, safety, and specific egg withdrawal times. Always check the label: If a product makes a claim, legally required data must support it. If that data is missing, you should question the product's validity. Rely on science, not emotionally driven marketing, to keep your flock healthy.

Parasite Management

Chicken parasites can be internal or external. Internal parasites include worms such as roundworms, tapeworms, cecal worms, gapeworms, capillary worms, and eye worms. Chickens can also be infected with protozoa such as coccidia. These ailments are common and normal, but can get out of hand quickly if you aren't paying close attention. Specific treatment for each type of parasite is found in chapter 6, but here are key principles to keep in mind.

- Use multiple layers of protection rather than relying on one tactic. Biosecurity, monitoring, coop maintenance, and treatments all play a role in managing parasites.
- Parasites thrive in moist, dirty, and overcrowded conditions. Your first line of defense is a solid maintenance routine and plenty of space for your chickens to spread their wings and forage on ample ground. If they are kept in a tiny coop and run, the poop and problems pile up fast. Rotate pastures where they forage to help break parasite cycles in the environment.
- Parasites can be managed fairly easily as long as you catch them in time. It's best to monitor droppings, behavior, feather condition, and overall wellness.
- Some chicken keepers need to deworm their chickens once or twice a year if their property has a heavy parasite load. Others can get away with treating only as needed.
- I suggest yearly fecal testing for parasites. Your vet can perform these tests for you, or you can send a sample to a lab.
- Use preventive measures to keep external parasites away. Check the Resources section at the back of the book for options.

Disease and Pathogen Management

Parasites are only one part of the equation when it comes to what could be making your chickens sick. They can get viral, bacterial, and fungal infections (as if we weren't worried enough). It can be difficult to prevent your chickens from becoming infected this way, but there are some essential factors to keep in mind.

- Viruses spread through the air, fecal-oral routes, wild birds, and contaminated equipment. Chickens can be carriers of the virus without showing signs, making containment difficult.

- Many viral diseases can be prevented with vaccines. Some must be given early, such as at hatch, while others can be given later. Not all vaccines prevent outbreaks, but they may lessen symptoms. They can't be treated with antibiotics, but there are remedies to lessen the effects.
- Bacterial infections can occur in tissues, fluids, or the environment and can enter a chicken through wounds, the respiratory tract, or the digestive system. Bacterial infections can be treated with antibiotics, but obtaining them without a veterinarian can be difficult. It is also irresponsible to overuse antibiotics without knowing what you are treating.
- Damp, moldy bedding can produce fungal spores. These spores are then ingested or inhaled, causing the chickens (or you) to become sick. Fungal infections are difficult to treat, so keep that coop clean, provide adequate ventilation, and protect yourself by wearing a mask or respirator during big cleanouts.

Over the years, I have learned how important it is to wear a mask or respirator when stirring up dust. Histoplasmosis and other illnesses can be serious problems.

VACCINATION CHART Here is a chart to help you navigate available vaccinations in the United States. For other countries, check with your national and local agricultural authorities. These are the vaccines commonly used by backyard chicken keepers. Others are available for commercial use but are not recommended for backyard use for various reasons.

DISEASE	PROTOCOL	NOTES
Marek's disease	Given to chicks at hatch by subcutaneous injection by owner or hatchery.	Marek's vaccination doesn't prevent the infection or viral shedding completely, but it can help avoid mortality. A flock that contracts Marek's will pass it on to birds added in later on, even if they survive.
Fowl pox	Given by puncture in the web of the wing at about 10 weeks old. Booster available before mosquito season.	Not as common. Recommended in areas with a history of fowl pox.
Newcastle disease	Given via eye drops, drinking water, or nasal drops when chicks are new, and then again around 2 to 4 weeks old. Boosters may be needed.	More practical for commercial operations because the vaccine comes in large vials. Usually only recommended in high-risk areas.
Coccidiosis	Spray or additive to feed or water in early chick stage.	Not to be given alongside premedicated feed as it negates the effect.

ASK THE POULTRY DOC
VACCINATIONS

A robust preventive medicine strategy is the cornerstone of a thriving flock, but because every farm is different, there is no one-size-fits-all protocol. Your strategy must be tailored to your specific geographic location, the local "field challenge" (disease pressure), your biosecurity capabilities, and your ultimate goals—whether that means raising organic layers or breeding show birds. For those raising chicks organically or wishing to avoid chemical coccidiostats like amprolium, vaccination against coccidiosis is a highly effective alternative that builds natural immunity from day one. Furthermore, flocks intended for breeding or long-term production often require expanded coverage against pathogens that impact fertility and longevity, including infectious laryngotracheitis (ILT), fowl pox, fowl cholera, avian encephalomyelitis (AE), and chicken anemia virus (CAV). Because navigating these variables can be complex, I recommend consulting with a poultry veterinarian to design a program that fits your specific risks. For a curated selection of veterinary-recommended preventive products and comprehensive care protocols, visit ThePoultryDoc.com.

TOXINS AND POISONS

CHICKENS THAT LIVE OUTDOORS, especially free-range birds, are exposed to hazards you should be aware of. Beyond predators, disease, and parasites, environmental toxins pose a serious risk if not managed. Many of these toxins can cause illness, organ damage, reduced immunity, contaminated eggs, and death. Awareness and precaution are essential to keeping your birds safe. You'll find the most common toxins below.

- **Spray drift** can enter the chicken area when nearby fields, yard, or gardens are sprayed, affecting insects, feed, bedding, chickens, and plants they eat. Keep chickens indoors and cover feed and water during spraying. Pollutants like fertilizers and herbicides can leach into soil, contaminate groundwater, and remain in puddles, which chickens may drink from.

This wisteria has been growing on my coop for years. It can be toxic, but because the chickens have plenty of foraging space, they have never bothered with it. Jasmine, morning glory, and some honeysuckles are great options if you want a vining plant that isn't toxic.

- **Toxic plants** can be everywhere, but chickens usually avoid them. Common ones include rhododendron, oleander, nightshade, foxglove, lily of the valley, hydrangea, rhubarb, and azalea. I have some in my yard, but the chickens stay away because of ample foraging space. If your yard is smaller, consider changing the landscaping.
- **Mold/mycotoxins:** Spoiled feed, damp bedding, and old substrate can harbor mold that produces mycotoxins, causing liver and kidney damage or immune issues. Exposure is dangerous for chickens. Store feed properly, keep feeders out of weather, and inspect for mold regularly.
- **Cedar shavings:** Bedding made from cedar or other aromatic wood can produce phenols that are irritating to the respiratory system.
- **Cleaners:** Some cleaning solutions are toxic to chickens. Follow all label instructions for dilution and allow the coop to air out completely before letting the chickens return.
- **Heavy metals:** Lead from old paint flakes, leaded gasoline, and metal scraps, as well as arsenic and other heavy metals, pose a risk to chickens. Avoid using these materials around the coop to limit ingestion. Don't allow scrap metal to accumulate and sit in the elements for long periods of time.
- **Air pollution:** Air pollution from backyard fires and wildfires can irritate respiratory systems, causing lung damage. Keep chickens indoors until the threat passes. Ash and soot can contaminate feed and water, so cover feeders and waterers when unused.

PECKING ORDER IN THE FLOCK

ONE OF THE BIGGEST MISTAKES chicken keepers make is ignoring the importance of maintaining a peaceful flock. Whether it's because they sense weakness or haven't been properly introduced, chickens will hurt each other. There will be pecking here and there, which is normal and part of the pecking order. If you see actual injuries or segregation from the flock, it is more serious. It is our job as chicken keepers to prevent bullying and stop it if it starts. You will find all my information on flock dynamics in chapter 9. Here are some basic tips:

- Integrate new birds slowly using the see-but-not-touch method. Separate chickens with fencing or crates so they can explore each other without causing injury.
- Add extra feeders and waterers when bringing in additional chickens. This reduces the amount they are territorial over.
- Remove bullying chickens from the flock for a "time-out." This gives the flock a break from the bullying and helps reset the pecking order.
- Be sure to provide plenty of space, enrichment, hiding spots, and good nutrition.

Threats to your flock lurk around every corner. Once you establish a routine, protecting them from these threats becomes second nature. Protection is a mindset of vigilance, layered defenses, and proper care. Combining strong structures with smart husbandry helps keep your flock happy and healthy. Next, I will go into detail about external ailments. You will find information for common ailments such as wounds, mites, lice, injuries, eye health, bumblefoot, and so much more.

My flock waiting for treats (which is their reward for not picking on each other)

5

COMMON EXTERNAL *Ailments*

The first time I found mites on my chickens really threw me for a loop. I was completely stressed about everything, from treatment to whether they would get inside my house and onto my family. It was a few days of unnecessary anxiety, and now, when my chickens get mites, I am calm and collected. When you encounter chicken ailments, you may want to panic, but instead, open this book and handle it like a professional. The following ailment profiles are based on my personal experience, observations, and research. It is meant to inform, not replace, professional veterinary care. If you ever face a case that is severe, rapidly progressing, or causing your chicken pain, please seek medical help from an avian or poultry vet.

The rest of this chapter contains ailment profiles for the most common external chicken issues. Many conditions show themselves first on the skin, feet, combs, feathers, or limbs. You may go years without experiencing any chicken ailments, but if you do, you can be prepared. This chapter should serve as a reference guide to help you recognize the early warning signs, provide immediate first aid steps, and know when to seek additional care.

On the following page is a quick flow chart to help you decide whether the ailment is something you can continue to treat at home or whether you will need help.

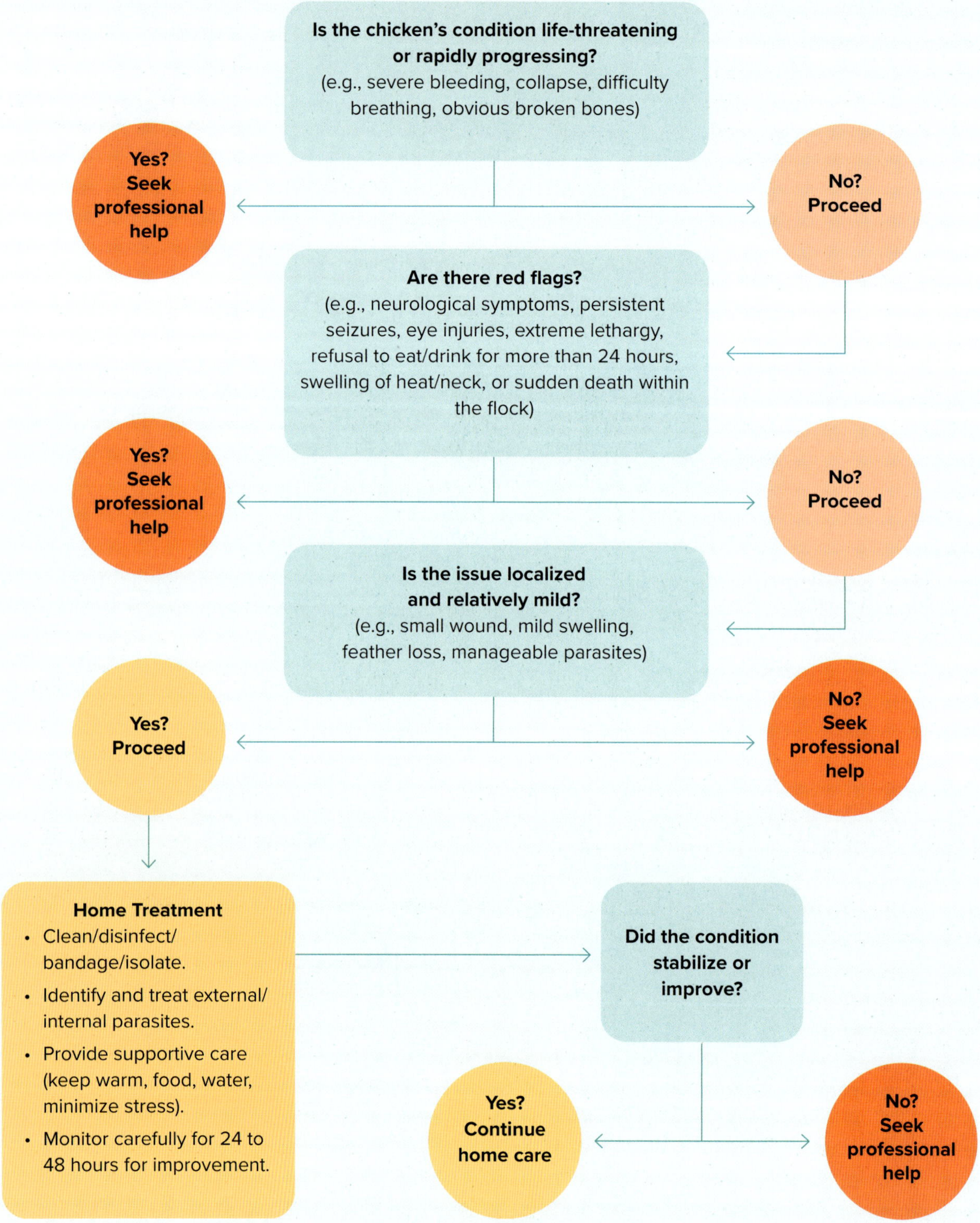
Is the chicken's condition life-threatening or rapidly progressing?
(e.g., severe bleeding, collapse, difficulty breathing, obvious broken bones)
Yes? Seek professional help
No? Proceed
Are there red flags?
(e.g., neurological symptoms, persistent seizures, eye injuries, extreme lethargy, refusal to eat/drink for more than 24 hours, swelling of heat/neck, or sudden death within the flock)
Yes? Seek professional help
No? Proceed
Is the issue localized and relatively mild?
(e.g., small wound, mild swelling, feather loss, manageable parasites)
Yes? Proceed
No? Seek professional help
Home Treatment
• Clean/disinfect/ bandage/isolate.
• Identify and treat external/ internal parasites.
• Provide supportive care (keep warm, food, water, minimize stress).
• Monitor carefully for 24 to 48 hours for improvement.
Did the condition stabilize or improve?
Yes? Continue home care
No? Seek professional help

MITES

The dust on this feather is actually mite eggs. The little black specks are the mites.

MITES ARE THE DREADED PARASITE for all chicken keepers! They are tiny external parasites that feed on blood. Northern fowl mites live on the bird but can also be found on the eggs, on the roosts, and in the nests. Red mites or roost mites can be found in the coop environment. Left unchecked, infestations can become quite heavy and can cause anemia, stress, stopped egg production, feather damage, and death.

Causes

- **Overcrowding:** Dirty conditions and poor ventilation can create an environment for mites to thrive and hide. Too many birds in one coop can cause mites to spread rapidly.
- **Lack of dust bath area:** Chickens rely on dust bathing to eliminate parasites and clean their feathers. Dust bathing areas should be available all year-round.
- **Weak birds:** Chickens who are dealing with illness or have a weakened immune system are less likely to care for themselves and contract mites.
- **Broody hens:** A hen who is sitting on a nest all day long due to being broody is more likely to attract mites because she isn't dust bathing and moving around.
- **Wild animals:** Birds and rodents can bring mites into the coop space.

Symptoms

- Increased preening, scratching, or restlessness.
- Scabs or redness around the vent or bare patches on the body.
- Mite eggs on the feathers that look like dust.
- Visible mites crawling on the skin.
- Dirty vent feathers.
- Unexplained weight loss, lethargy, or lack of egg production.
- Pale combs or wattles indicating anemia or stress.

Diagnosis

Mites hide during the day, so they are easier to find at night. Remove chickens from the roosts at night. Use a flashlight or headlamp to inspect the skin under the feathers, focusing on the vent area and under the wings. Mites are smaller than a pinhead and move around on the skin. During the day, you may see mites on the chicken or crawling on your hands after handling them. You can also search the feathers for eggs. Run your finger across the roosts early in the morning. Well-fed mites will leave a streak of blood on the roost.

Treatment

There are several options for treating mites, and your preferred method will depend on convenience, experience, and cost. Be sure to treat both the birds and the entire coop to prevent reinfection.

METHOD 1: Approved Chemical Treatments

Use a poultry-safe product such as spinosad (Elector PSP or Fraxel PSP) or permethrin to treat the flock. Follow all label instructions and store properly for next time. My preferred method is Spinosad products because they kill adult mites and their eggs, so you do not need to repeat the process. There is also no egg withdrawal time when used properly.

METHOD 2: Off-Label Products

Off-label products, such as Ivermectin, macrolides, or lactones, are often recommended by chicken keepers to treat mites. These products are intended for other species and have not been tested on chickens. Chicken keepers often choose them because they are cheaper up front. However, you usually need to treat more than once, and the use itself is risky because you have to rely on an unapproved dosage. Please consult a vet before use and pay attention to egg withdrawal recommendations.

METHOD 3: Dusting

Dusting the flock with food-grade diatomaceous earth can deter mites, but it must be applied liberally and repeatedly for it to be effective. Even food-grade DE can cause health concerns for you and your chickens, so be cautious if you choose this method.

METHOD 4: Homemade Natural Remedies

Dish soap and essential oils can kill mites on contact. The treatment needs to be repeated several times to be effective. Mix a few drops of dish soap, 10 drops of cedarwood essential oil, 10 drops of lavender essential oil, and 1 ounce (30 ml) of fresh-pressed garlic juice in 16 ounces (475 ml) of water. Shake well, then spray the chickens' skin, focusing on the vent and under the wings. Repeat every other day for up to 2 weeks. Watch for skin sensitivities. Clean out the coop and spray every crack and crevice liberally. Repeat as needed.

Prevention

- Inspect your flock regularly, especially during warm, damp, or humid months. I notice mites most often in the spring and fall.
- Maintain a clean coop with minimal clutter. Remove fallen feathers during molting season. Keep rodents and wild birds out of the coop and run.
- Avoid overcrowding and give chickens space to move around freely.
- Provide a dust-bathing area all year.
- Use a preventive dust such as the one listed in the Resources section.

When treating a large flock for mites, I like to use a tub to make sure they are fully covered in the treatment. Swish them around to be sure it gets under the feathers and onto the skin.

SCALEY LEG MITES

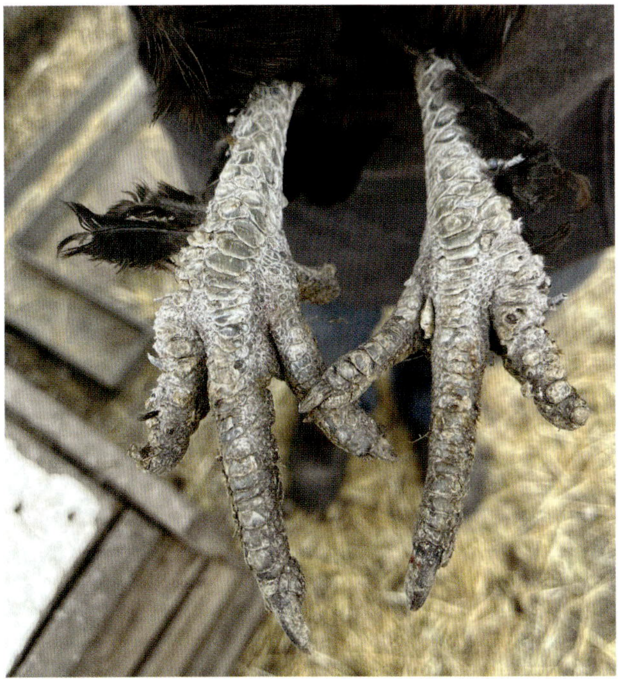

This is an image from a social media follower. They were kind enough to rescue a chicken that was kept on a gravel floor with unsanitary conditions, leading to scaley leg mites.

KNEMIDOCOPTES MUTANS are tiny mites that burrow beneath the scales of a chicken's legs and feet. They feed on dead tissue, causing lifted, hardened scales and irritation. The infestation worsens over time, leading to deformation and extreme discomfort.

Causes

- Direct contact with other infected chickens or wild birds.
- Established colonies will persist in the environment unless properly controlled.
- Cracked roosting bars or unsanitary coops.
- Lack of monitoring allows infestations to take hold.

Symptoms

- Feet or leg scales that are lifted, crusted, rough, and thickened.
- Uneven scales with gaps.
- Discomfort, lameness, stiffness, or swelling of the legs.
- Deformation or necrosis of the toes.

Diagnosis

Visually inspect the legs and feet regularly to prevent an infestation from getting out of hand. As chickens age, their leg scales can become rougher, so it is important to distinguish normal aging from scaly leg mites. Legs infested with mites will worsen over time, becoming thicker and crustier. If possible, use a microscope to identify the mites in a scraping taken from under the scales.

Treatment

1. Soak the chicken's legs and feet in a few gallons of warm water with 1 cup (240 g) of Epsom salt. Soak for 15 to 20 minutes.
2. Very gently scrub the legs to remove loose scales. Be careful not to tear the scales and cause bleeding.
3. Dry the legs and apply vegetable oil with one drop each of lavender and frankincense essential oils. For extreme cases, consider adding a sulfur dip before applying the oil.
4. Cover the legs with a barrier such as a shea butter or petroleum jelly to suffocate the mites.
5. Repeat the process every few days until the legs are back to normal.

Prevention

- Provide a clean coop and run with space for dust bathing.
- Avoid introducing new flock members until after a quarantine period.
- Keep rodents and wild birds out of the enclosure.
- Inspect legs and feet often for early detection.

POULTRY LICE

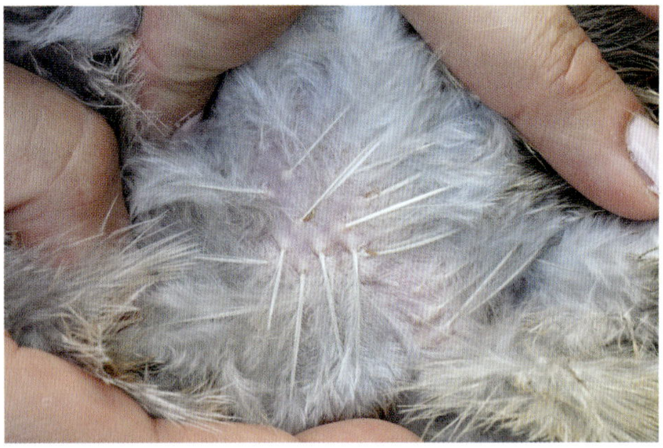

Mandy came to me as an adult, and when I brought her home, I found lice crawling on her skin! Fortunately, I quarantined her from my flock and was able to treat her.

POULTRY LICE ARE EXTERNAL parasites that feed on feathers, skin debris, and sometimes blood. Lice spend their entire life cycle on the bird and usually don't infest the whole coop like mites do. Infestations start mild but quickly worsen, leading to feather loss, anemia, and even death.

Causes

- Lice can be introduced to the flock through the addition of new birds, infected equipment or crates, or wild birds.
- Overcrowding and poor hygiene tend to make chickens more vulnerable.
- A lack of dust-bathing spaces can make feathers more prone to infestations.

Symptoms

- Light brown lice are found crawling on the skin or feathers of the chicken.
- Visible eggs are found on feather shafts that are usually concentrated at the base. Eggs are harder to find until the infestation is heavy.
- Increased preening, scratching, and restlessness.
- Feathers can be dull, broken, or missing. Wattles and combs may appear pale, especially if the infestation is advanced.
- You may notice reduced appetite, weight loss, and a drop in egg production. Extreme cases will cause lethargy.

Diagnosis

To check for lice, part your chicken's feathers and look for moving lice or eggs on the feather shafts. Eggs are often light gray and clustered; if you see lice but no eggs, your chicken is infested. Use a magnifying glass, focusing on the vent, under the wings, and on the chest. If one has lice, the whole flock likely does. Lice won't transfer to you.

Treatment

1. When introducing a new bird, isolate it from the flock and observe for lice for the first few weeks.
2. Use an insecticidal treatment labeled for poultry lice, like spinosad products such as Elector PSP or Fraxel PSP, even if labeled for mites. They work well and kill adults and eggs without an egg withdrawal period, making one treatment effective. Other options include permethrin dusts or sprays. A second treatment in 10 days is necessary to kill hatchlings. Follow product label instructions.
3. Clean the coop by removing all bedding, scrubbing surfaces, and spraying everything with your desired product. Allow the coop to dry before adding bedding and chickens.
4. Support any lethargic or weak birds by providing them with good nutrition, clean water, and vitamins.

Prevention

- Quarantine new birds for 2 to 4 weeks before adding them to your flock.
- Inspect the flock on a regular basis. I suggest a monthly feather inspection.
- Eliminate overcrowding and provide a clean space for your flock.
- Provide dust-bathing areas year-round.
- Keep your flock separate from other flocks and take biosecurity precautions when visiting other farms or homes with birds.

STICKTIGHT FLEAS

The chick on the left has a bad case of sticktight fleas while the hen on the right has scabs from pecking wounds. It is easy to confuse the two without looking closely.

THIS EXTERNAL PARASITE embeds itself in the chicken's skin. You'll typically find them on combs, wattles, or around the eyes of the chicken. The flea stays attached to the skin to feed on blood and lay eggs that drop onto the chicken or the environment. The biggest concern is anemia from blood loss, so the infestation should be addressed promptly.

Causes

- Fleas can get into the bedding, litter, or free-range space due to wild birds, rodents, or other animals. They prefer rich soil or straw bedding.
- Limited dust-bathing areas, bare patches on the body, or thin feathering make the bird more vulnerable.
- Warm, humid weather can cause a greater risk.

Symptoms

- Sticktight fleas will show up as specks on the comb, wattle, or around the eyes.
- Chickens will scratch at their faces more often due to irritation.
- The chickens may have pale combs and wattles, swollen eyelids, or general weakness.
- You may see reduced egg production.

Diagnosis

Inspect your birds regularly, especially the skin around their face, for dark specks that are often in clusters. These may be fleas or scabs; hold a wet towel on the speck to see if blood shows, indicating a scab. A magnifying glass can help as well.

Treatment

1. Remove adult fleas from the bird using tweezers. Place the removed fleas in a cup of soapy water. Apply antiseptic or antibiotic ointment to the area to prevent infection.
2. Use a powder or spray treatment to spot treat the bird. Permethrin or spinosad are usually the preferred products, just as with lice and mites.
3. Clean out the coop completely and treat the floors, nesting boxes, corners, and gaps. If possible, clean out the run substrate and spray the ground. If you have dry sand in your run, you can mix in a treatment while the chickens are free-ranging and let it settle. Burn any bedding you have removed and allow the space to dry before adding new bedding.
4. Repeat treatments every 10 days until the fleas are eliminated.

Prevention

- Maintain a clean and clutter-free environment.
- Add a roof to the run to keep areas dry.
- Limit access to wild animals.
- Isolate new birds before adding them to your flock.

BUMBLEFOOT

A follower sent me this image of her chicken's bumblefoot. You can see the infection is causing the foot to swell.

BUMBLEFOOT IS A FAIRLY COMMON footpad infection among backyard chickens. It begins as a small puncture, cut, or injury to the bottom of the foot and progresses into a deeper abscess. Most of the time, Staphylococcus is the bacterium responsible. Unfortunately, it can become quite serious and must be addressed as soon as possible to prevent infection of the tendon, bones, or bloodstream. Not only is bumblefoot painful, but it can also cause lameness or death if left untreated.

Causes

- Bacteria enter the footpad through a cut, scrape, splinter, or puncture wound.
- Roosts that are too high can cause heavy chickens to get injured from jumping to the floor.
- Hard or rough surfaces in the coop or run can cause injuries.
- Wet, muddy, or soiled coops and runs can weaken the foot skin and leave the chicken more susceptible to injury.
- Nutritional deficiencies and poor health can increase the likelihood of skin injury.
- Small roosts can put too much pressure on the footpad, especially in heavier breeds, leading to open skin sores that can become infected.

Symptoms

- Limping or favoring one foot, reduced activity.
- Reluctance to jump up to roosts. They may prefer sleeping on the floor.
- Swollen foot, especially on the footpad, redness or heat in the area. Swelling may appear on the top of the foot if the infection is deep enough.
- Dark-colored scab on the sole of the foot or the toe pad.
- If infection spreads, the chicken may seem unwell, lose weight, or stop egg production.

Diagnosis

Physically examine your chicken by wrapping it in a towel or holding it securely. I prefer to wash the chicken's foot well to get a better look at what is going on. Sometimes your chicken will get a scab on the bottom of their food from an injury, and it won't become infected. If it looks like it is healing fine on its own, leave it alone. If there is swelling, redness, or the scab isn't healing well, the bumblefoot needs to be treated. Bruising from a high jump, gout, and other joint problems can resemble bumblefoot swelling. Be sure to look for the redness and scab to confirm you are dealing with bumblefoot. In advanced cases, a veterinarian may need to X-ray the foot to check for bone or tendon involvement. They can also culture the bacteria to determine whether an antibiotic is necessary and offer pain management. It is important to inspect your flock's feet often because catching it early makes it easier to treat.

Treatment

EARLY-STAGE BUMBLEFOOT

1. After washing your chicken's feet, fill a bucket with enough warm water to cover the feet and legs of your chicken. Add about ½ cup (120 g) of Epsom salt per 1 gallon (3.8 L) of water, then let it dissolve.
2. Soak the feet for 20 minutes or until the footpads soften.
3. Apply an antiseptic and allow it to dry. Follow with an antibiotic ointment and gauze. Wrap the foot with self-sticking vet wrap.
4. Reapply the antibiotic ointment and wrap the foot every other day until healed. It is best to keep the bird in a cage until the foot is healed to prevent jumping from high places.

MODERATE BUMBLEFOOT

1. Complete the steps above, but use a drawing salve to push the infection to the surface. It may take 7 to 10 days of reapplying and wrapping, but the infection should surface and make for easy removal by you or a veterinarian.
2. Once it has surfaced, gentle pressure could be enough to push out the hard kernel of infection and the surrounding pus. If not, a scalpel may need to be used to create a small incision. Chickens don't always show signs when they are in pain, but cutting into their skin is painful and should be done by a vet who has access to pain management.
3. Once the kernel has been removed, the wound should be flushed with saline or a very dilute chlorhexidine solution.
4. Apply antibiotic ointment and gauze before wrapping with self-sticking vet wrap. Redo this application daily until healed. Keep your chicken isolated so they don't further injure the area.

Prevention

- Keep roosting bars at a moderate height with ramps to help the chickens get down.
- Ensure your flock gets balanced nutrition with correct vitamins, minerals, and protein to support skin and immune health.
- Control treats and scraps to prevent chickens from becoming overweight.
- Maintain good coop hygiene and provide soft bedding.
- Remove abrasive surfaces, sharp edges, or anything that could potentially cut the foot area.

A NOTE ABOUT FOOTPAD DERMATITIS (FPD)

Footpad dermatitis (FPD) is a contact dermatitis caused by prolonged exposure to damp, ammonia-rich bedding. The foot can become inflamed and discolored, with ulcerative lesions on the bottom of the foot. Bumblefoot is different, as it refers to a deeper bacterial infection and abscess. To prevent FPD, keep the coop and run clean and dry.

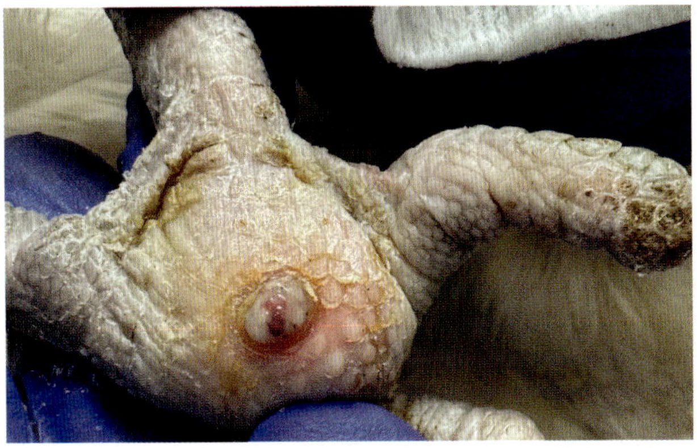

This is the footpad with a sore, which is where the infection started.

BIG CONSEQUENCES If bumblefoot goes undetected, it can become quite severe. Not only does the footpad become inflamed and painful, but the infection can also set deeper into the tissues, bones, and tendons. It is important to seek out veterinary care for surgical debridement, pain management, and proper antibiotics.

DEPLUMING MITES

Depluming mites

DEPLUMING MITES tunnel into the skin at the base of the feather shafts, so they are hard to see, leading many to believe their flock is mite-free. They feed not only on the skin and feathers but also on the fluids produced by the damaged skin. Chickens often develop bare patches and may pluck their own feathers or those of others.

Causes

- Transmission occurs when the flock is in close contact with infected chickens or wild birds.
- Chickens that are stressed, crowded, or sick may be more susceptible.
- Weak or stressed chickens are more likely to suffer a heavier outbreak.

Symptoms

- Thin feather patches and bare skin, particularly on the back, around the neck, on the wings, or near the vent.
- Broken and tattered-looking feathers with unexplained reasons.
- Reduced or halted egg production.
- Red, irritated, and inflamed skin around bare patches.
- Weight loss and poor body composition.

Diagnosis

Diagnosis involves observation, inspection, and elimination of all other causes. Bare patches and broken feathers can also result from rooster mating behavior or feather picking. Inspect the broken feathers. Often, the mites hollow out the shafts, weakening them. A veterinarian can identify the mites through a skin scraping or a feather sample.

Treatment

1. Treat the birds with a sulfur powder dusting or an acaricidal spray. Apply the product as close to the skin as possible, focusing on the underwings and the base of the tail. Wear a mask while you work. Ivermectin is another treatment option, but it is off-label, so be sure to consult a vet for dosage and application.
2. Completely clean the coop by removing bedding, dust, and nesting material. Apply the chosen mite product on the coop surfaces before adding fresh bedding.
3. Freshen their dust-bathing areas.
4. Complete a second treatment and coop cleaning after about 2 weeks to catch any newly hatched mites.
5. Feathers may not grow back right away, but should return after a fall molt. Be sure to provide the flock with a higher protein feed during this time.

Prevention

- Maintain a clean coop by refreshing bedding and nesting materials.
- Keep wild birds away by sealing the coop and run.
- Don't purchase new chickens from places with adult chickens who are missing feathers. Quarantine new chickens before adding them to your flock.

EYE INFECTIONS

This hen belongs to one of my followers. Eye infections can quickly get out of hand. Luckily, she had someone caring for her!

EYE INFECTIONS IN CHICKENS are conjunctivitis or ocular inflammation involving the membranes around the eyes, eyelids, or deeper tissues. They can occur in one or both eyes and can be contagious. An eye infection is not only uncomfortable for a chicken, but it can also affect vision, eating behavior, and overall health.

Causes

- Foreign bodies such as dust, straw, sticks, feather fragments, or the beak of another chicken can scratch the eye.
- Chemical exposure, such as ammonia, dust, or cleaning solutions, can cause the eye to become inflamed and start an infection.
- Chickens often scratch their eyes when irritants, mites, or sticktight fleas are present. Excessive scratching by the chicken can lead to infection.
- Eye worms, a parasite that lays eggs and lives in the eye area, can cause scratching or infection.
- Poor ventilation, high ammonia, and dirty conditions can weaken the protective mucous membranes of the eyes, which can lead to infection.

Symptoms

- Eye discharge or crusting around the eyes.
- Redness and inflammation in or around the eyes.
- Swelling on the face, eyelids, or eye.
- Excessive scratching of the eye area or keeping the eyes closed.

Diagnosis

Chickens are good at hiding ailments. Be sure to inspect your flock regularly. Eye infections alone won't cause respiratory symptoms, but they may have a respiratory illness that led to an eye infection. If there are signs of respiratory distress, such as wheezing, coughing, sneezing, eye watering, facial swelling, or nasal discharge, the respiratory illness will need to be treated separately.

Treatment

1. Isolate the chicken to reduce the spread of the infection.
2. Flush the eye with sterile saline to remove the crust, discharge, and potential eye irritants. Clean around the area with a warm, damp cloth.
3. Apply a topical antibiotic ointment or drops as directed on the label. If you are unable to locate such medicines, veterinary guidance might be needed.
4. If needed, clean the coop and address common factors such as dust, ventilation, bedding, and cleanliness.

Prevention

- Keep bedding dry, remove manure buildup, and prevent dust and ammonia with good ventilation.
- Prevent squabbles and pecking by providing ample space and entertainment for the flock.
- Monitor and treat for parasites as needed.
- Quarantine new flock members to monitor for early signs of disease.

FROSTBITE

Big Mike suffers from a bit of frostbite on his wattles and comb.

ONE OF THE HARDEST THINGS about winter is frostbite. Chickens are great at regulating their systems to cope with the cold, but frostbite can still affect the feet, wattles, and combs when these areas are exposed. It occurs when skin is exposed to the cold for too long, causing ice crystals to form in the cells. This compromises circulation, leading to necrosis (tissue death), infection, pain, and suffering.

Causes

- Prolonged exposure to freezing temperatures.
- Freezing temperatures paired with a lack of protection from wind, ice, snow, and wet conditions.
- Excessive moisture and humidity in the coop.
- Wet wattles from dipping them in open watering systems.
- Wet feet and combs from damp conditions.
- Thin roosts that don't allow the chicken to lay their feet flat and cover them with their body.
- Frozen ground without a dry space to retreat to.

Symptoms

- White, gray, or dull yellow tissue may indicate the onset of frostbite. Blue, gray, or black discoloration is a sign of necrosis.
- Swelling, hard areas of skin, or blisters may appear.
- Lameness, weakness, reduced appetite, reduced egg laying, or general discomfort.
- Severe cases can lead to dead tissue falling off.

Treatment

1. Immediately move the bird to a warm, dry, draft-free location. I usually use my garage with a space heater or radiant heater, but you could also bring the chicken into the house. Once the chicken is in a warm location, do not expose it to cold until it is fully healed, so prepare a space you can use long term.
2. Allow the areas to gradually warm. You can use lukewarm water to soak their feet or use it on a cloth to gently warm combs and wattles. Do not rub, massage, or apply pressure. Avoid direct heat from a blow dryer, heat lamp, or other extreme heat source.
3. Apply an antiseptic or wound care spray.
4. Monitor for signs of infection, such as pus, swelling, and odor. Consult a veterinarian if this happens.
5. If dead tissue begins to fall off, do not remove it yourself. Allow it to heal naturally and continue using the antiseptic spray and antibiotic ointment if needed.

SUNBURN

Did you know chickens can get sunburn? While uncommon, it can still happen! The best way to prevent sunburn is to provide shade for your flock. You can also use chicken saddles to cover their skin when they are missing feathers.

Prevention

- Insulate your coop and prevent drafts by covering gaps around or below the roost height. Keep ventilation high in the coop to allow moist air to escape.
- Use flat roosting bars to allow the flock to cover their feet.
- Keep moisture down in the coop by having ventilation up high, keeping waterers outside the coop, and reducing soiled bedding.
- Give the flock a protected space during the day by wrapping the run in clear plastic and installing a roof.
- Choose cold-hardy breeds if you live in a location that gets below freezing. Avoid chickens with large comb styles.
- Focus on your chickens' health with high-quality feed and nutrients so they are prepared to regulate body temperature and maintain heat.

ASK THE POULTRY DOC

USING A PREVENTIVE CREAM

While it is a common practice to coat combs and wattles in petroleum jelly or thick salves as a "shield" against the cold, I generally advise against this strategy. The primary enemy in frostbite is moisture, not just low temperatures. Heavy ointments can inadvertently trap moisture against the skin, actually increasing the risk of freezing rather than preventing it. Furthermore, these sticky substances often rub off onto the bird's feathers, matting them down and destroying the insulating air pockets they rely on for warmth. Instead of applying topical barriers, the most effective prevention is environmental, specifically, the use of supplemental heat. Keeping the coop temperature above critical lows provides a dual benefit: It protects tissue directly and significantly reduces relative humidity. This combination—warmer air and lower moisture—compounds to form the single best defense against frostbite.

LACERATIONS

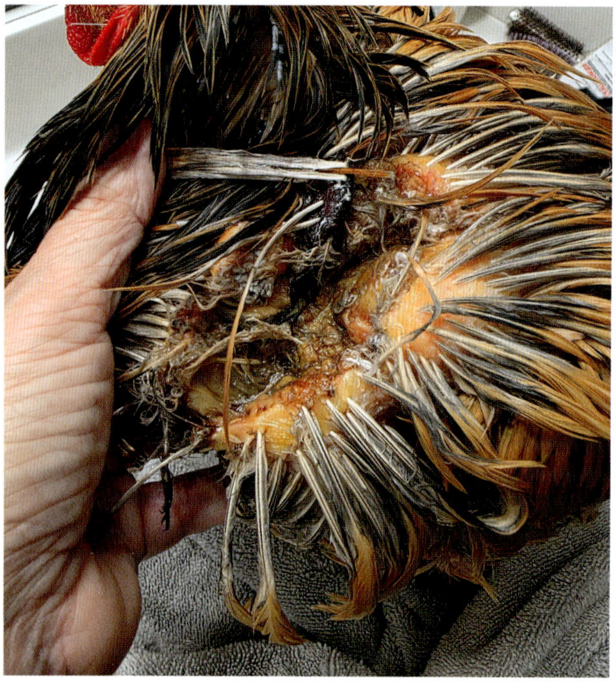

A follower sent a photo of her chicken attacked by a fox. The laceration was hidden under feathers, going unnoticed for days. Chickens can heal surprisingly well, and with patience and care, this girl pulled through.

LACERATIONS ARE CUTS OR TEARS in a chicken's skin. They can be superficial or deep, exposing or cutting through underlying tissue. These scrapes or large, gaping wounds can cause bleeding, infection, or even death. Open wounds can lead to pecking by the other chickens, so they must be handled quickly. Chickens can surprisingly heal from some scary-looking wounds, so don't be discouraged.

Causes

- Attacks from predators, other chickens, or even domestic pets.
- Sharp edges on coop or fencing structures.
- Mating with a rooster with sharp spurs.

Symptoms

- Visible cut or tear to the skin.
- Bleeding or dried blood on the wound or surrounding feathers.
- Limping, favoring one limb, avoiding contact with the other chickens, and a change in normal behaviors.
- Foul odor, discharge, feathers sticking together, or maggots in or around the area (please refer to Fly-Strike if maggots are present).

Treatment

1. Remove the chicken from the flock to prevent added stress and pecking.
2. Control bleeding with pressure using a sterile gauze or a clean cloth.
3. Remove debris from the gash by gently flushing the wound with saline. A diluted chlorhexidine solution may be used to disinfect the area. Trim feathers as needed so they don't dry into the wound during healing.
4. Apply a poultry-safe antiseptic spray followed by an antibiotic ointment. If the chicken is able to pick at the wound, use gauze and vet wrap to cover the area.
5. Provide the bird with electrolytes and good nutrition during this stressful time. Scrambled eggs may encourage eating.
6. Check the wound daily for infection and redress as needed. If signs of infection are present, consult a vet for possible antibiotics.
7. Only return the chicken to the flock after it is fully healed. If blood, scabs, or wounds are present, the others will likely peck at the area.

Prevention

- Trim the rooster's spurs regularly.
- Check coop and run structures for protruding screws, nails, or wire fencing.
- Secure the chickens from predators in a safe coop a and run.

ASK THE POULTRY DOC

VETERINARIAN VS. AT-HOME CARE

Deciding whether a chicken laceration needs stitches (or more advanced care) depends on timing, contamination, and severity. If the cut is fresh, clean, and the edges come together easily, closure may be an option; if it's older (found late), very dirty/contaminated, or has missing tissue/crushed skin, it's often safer to manage it as an open wound with good cleaning and supportive care rather than "sealing in" bacteria. Chickens are resilient, and even significant skin injuries can heal surprisingly well with supportive care alone (clean housing, limiting pecking, stable nutrition, and monitoring). Get a vet involved promptly if you see any of the following:

- Gaping wound where skin edges won't appose, or a flap with questionable blood supply.
- Ongoing bleeding, expanding inflammation/redness, or signs of shock/weakness.
- Bite wounds, punctures, or heavily soiled wounds (manure, dirt, deep bedding).
- Exposed bone/tendon or involvement of joints/feet, face/eye, vent, or crop region.
- Bad odor, heat, worsening redness, maggots, or the bird stops using the limb.

Any surgical closure should be done by a veterinary professional—ideally with pain control and antibiotics only if warranted. For topical support, medical-grade honey products (including Manuka-type honeys) may be a helpful adjunct to keep the wound bed moist and discourage bacterial overgrowth, but they're not a substitute for proper antiseptic cleaning, debridement when needed, and appropriate veterinary care.

FLY-STRIKE

Fly-strike is pretty gross.

MOST EXPERIENCED CHICKEN KEEPERS would agree that fly-strike is one of the worst ailments to deal with. Fly-strike occurs when a wound goes unnoticed and attracts flies, which then lay eggs in and around the wound. Those eggs quickly hatch into larvae that feed on the bird's tissue. These maggots rapidly cause the chicken's health to decline, leading to severe discomfort and death.

Causes

- If a chicken has wounds, sores, or cuts, then flies may lay eggs there.
- Vent gleet or feathers with caked-on droppings or diarrhea will attract flies.
- Coops that are unkempt can attract more flies, especially at night when the chicken is sleeping.
- Weak chickens are less likely to dust bathe and preen, which leaves them more susceptible to fly-strike.

Symptoms

- Foul odor from wounds or the vent area.
- Visible maggots crawling anywhere on your chicken.
- Lethargy, self-isolation, or a change in eating or drinking habits.
- Matted or soiled vent feathers.

Treatment

1. Thoroughly clean the affected area by gently giving your chicken a bath with soapy water or gently flushing with saline.
2. Remove all maggots using tweezers, patience, and a strong stomach. Good lighting, a magnifying glass, and a helper may be needed.
3. Disinfect the area with a poultry-safe antiseptic or diluted chlorhexidine solution. Dress the wounds as needed.
4. In more severe cases, a veterinarian may need to surgically remove damaged tissue or provide antibiotics.
5. Keep the chicken in your infirmary for supportive care until healed.

Prevention

- Minimize the fly population by keeping the coop and run clean, removing soiled bedding and standing water, and moving compost or decaying matter away from your flock.
- Monitor chicken health and behavior. Check under the feathers for hidden lacerations or wounds when a chicken is acting off.
- Treat any skin damage right away to prevent attracting flies.

LEG INJURIES (SPRAIN, SLIPPED TENDON, FRACTURE)

LEG INJURIES IN CHICKENS encompass a wide range of problems, including sprains, strains, muscle damage, slipped tendons, fractures, and more. These injuries affect mobility and lead to other health concerns. When a chicken is unable to access their food, water, and dust bath, their health and egg-laying ability will suffer. Most chickens live outside and are susceptible to leg injuries from jumping, running, roosting, predators, and more. As humans, we rest, elevate, and ice our leg injuries. Chickens try to hide their injuries and will keep using the leg to the best of their ability, which can further injure it.

Causes

- High jumps or falls from roosting bars, perches, tree branches, and other structures.
- Flying into a building or wire mesh when frightened.
- Predator attacks.
- Fights among the flock.
- Wet or uneven flooring can cause a bird to slip and twist a leg.
- Nutritional deficiencies leading to bone weakness.
- Predisposition due to congenital disorders or malformation.

Fractures/Broken Bones

SYMPTOMS

- Leg is held at an odd angle.
- Serious breaks may result in a bone protruding from the skin.
- Chicken won't bear weight on one leg.
- Limping, dragging, or squatting often.
- Swelling, discoloration, or green-colored bruising.

TREATMENT

1. Isolate the bird in a quiet space. The isolation cage shouldn't have anything to jump from. Assess the injury to determine whether the bone might be broken. Some fractures are impossible to tell just by looking and would require an X-ray.
2. Contact the vet if possible, as it is best to have a vet set and stabilize the leg. For simple fractures that aren't near a joint or protruding through the skin, you can attempt this on your own.
3. Use a tongue depressor to give the leg stabilization. Cushion it against the skin with gauze padding and wrap with self-sticking vet wrap. Secure with medical tape.
4. Check often for swelling, discoloration, coldness, or other signs of compromised circulation.
5. Keep the chicken isolated and confined until the leg has fully healed. The healing process for a simple fracture is about 2 to 3 weeks.

Sprains, Strains, Torn Tendons or Ligaments

SYMPTOMS

- Limping, favoring one leg, no obvious break.
- Swollen joint, warm to the touch.
- The leg appears shaky or nonsupportive.
- Mobility is reduced, the chicken stays away from flock members, and sits often.

TREATMENT

1. Isolate the chicken in a confined area where they can rest and recover.
2. Gently palpate the area to identify swelling, heat, or abnormalities. Use vet wrap to support the injury if located.
3. If the injury doesn't appear severe, give the chicken time to rest in a crate that limits movement. Give the chicken food, water, and soft bedding. For torn tendons and ligaments, a sling or wrap to keep the leg immobilized is the best chance for healing.
4. A vet should be used to assess severe injuries. They may provide pain and anti-inflammatory medications.
5. If you do not see improvement after a few weeks, consult with a veterinarian.

Soft-Tissue Injuries/Bruising

SYMPTOMS

- Chickens who experience bruising on the leg areas, but don't have any signs of breaks, fractures, or sprains, may limp or favor that leg.
- There may be swelling.

TREATMENT

- Allow the chicken time to rest in a safe space.

Prevention

- Choose roosting bar heights that are appropriate for your flock. If you have heavier birds, the roosts need to be lower. I prefer my roosts to be a little above waist height.
- Provide ramps to access and exit the roosting area.
- Prevent slippery surfaces in the coop and run.
- Use a thick layer of soft bedding in case a chicken decides to jump down from the roost.
- Protect your flock from predators as best you can.
- Allow for ample space for your flock size.

PRO TIP

Did you know that a chicken bruise can be a few colors, but it is usually green?! This bruise is 3 days old on this chicken's neck. The bright green coloring means she is dealing with a painful bruise!

WING INJURIES

A chicken's wing is so beautiful! The intricate patterns are a natural work of art.

WING INJURIES IN CHICKENS can range from sprains and bruises to dislocations and fractures of the wing bone. They can affect the bird's mobility and welfare, so they should be treated as soon as they are discovered. I have seen wing injuries more often in adventurous, flighty breeds than in larger breeds, which tend to stay on the ground.

Causes

- Trauma from predator attacks.
- Collisions with coop walls, fencing, low branches, or run walls. This is usually due to a chicken being spooked into taking flight.
- Flapping wings in confined spaces.
- Squabbles with flock mates.

Symptoms

- A wing that hangs down loosely and is visibly lower than the other.
- The chicken may be struggling to roost or balance and may avoid jumping up to perches.
- You may notice swelling, bruising, or abnormal angles of the wing.
- The chicken may walk awkwardly or be reluctant to move around.

Treatment

1. Remove the chicken from the flock and take them to a stress-free space to investigate. Assess the injury by carefully examining the wing.
2. If you suspect a broken bone or notice a protruding bone, seek help from your veterinarian. If not, you may proceed with care for a sprain or other simple injury.
3. Immobilize the wing by folding it into the natural resting position. Once the wing is comfortably against the chicken's body, use self-sticking vet wrap to hold it in place. Go around the chicken's body and injured wing with the vet wrap, but leave the other wing out of the wrap.
4. Limit movement by keeping the chicken in a safe space, such as a crate, without perches or platforms for jumping. Give the chicken food, water with electrolytes, and soft bedding.
5. After 3 days, unwrap the wing and assess the injury. Rewrap if needed. If the wing appears healed, leave the wrap off, but keep the bird in the safe space for a few days to make sure it is fully healed.

Prevention

- Provide sufficient space in the coop and run for jumping and flapping.
- Be sure to hang roosts at a safe height and provide a ramp for entering and exiting the area. Some chickens may still want to jump down to the floor, so leave enough room in the coop for landing.
- Protect your flock from predators and maintain a balanced pecking order by properly integrating new birds and avoiding overcrowding in the coop.

BEAK INJURY

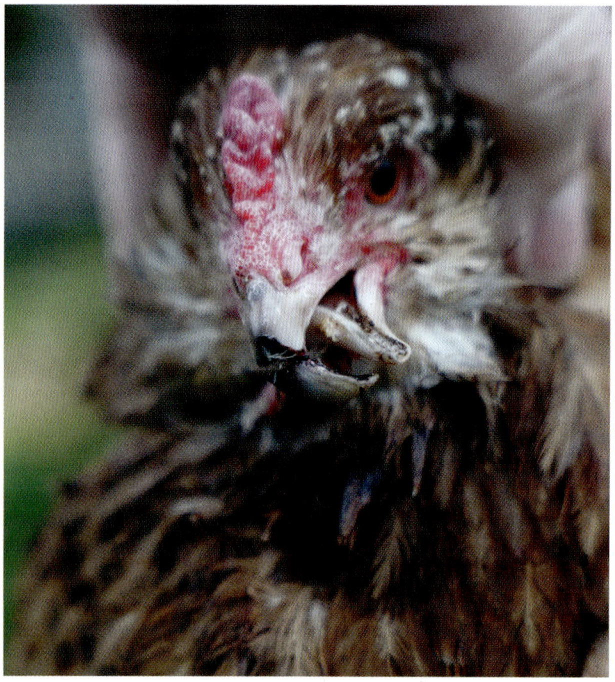

CiCi, my cross beak hen, broke the tip of her beak while foraging in gravel.

CHICKENS USE THEIR BEAKS for exploring, eating, drinking, scratching, preening, and interacting with the environment. It is one of the most important tools on their body, so if it becomes injured, it can be detrimental. The beak is made of sensitive tissue and bone with a hard keratin coating. If broken, chipped, or cracked, the area can be very painful and interfere with eating, drinking, and foraging. Minor injuries can resolve on their own, but more severe injuries need to be addressed to prevent infection and pain.

Causes

- Collisions against fencing or walls after being spooked or trying to fly in low clearance areas.
- Predator attacks.
- Crowded conditions that lead to fighting, bullying, or insufficient space to move safely.
- Getting the beak caught in fencing, wire mesh, or gaps in the coop structure.
- Rough foraging, such as looking for insects and food under gravel areas.

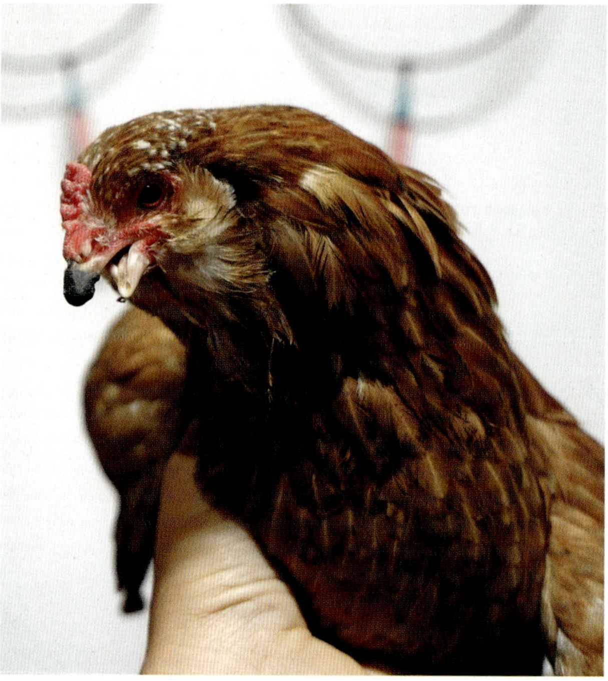

The vet was able to remove the broken piece, stop the bleeding, disinfect the wound, and create an epoxy tip to protect it while it healed.

Minor Crack or Chip

Many minor cracks or small chips will heal on their own. Beaks continue to grow, like a fingernail, and the chicken will file it down as it grows. This eliminates the crack or chip over time.

TREATMENT

1. Assess the injury to be sure it is superficial and can be left alone. If there is no bleeding or hanging pieces and the crack or chip isn't high up, closer to the face, the beak can likely be left to heal.
2. Offer softened food and fresh, clean water while the beak heals.

Cracked Beak with Keratin Layer Damage

If the crack or chip is deeper and affecting the keratin layer of the beak, you may notice bleeding, exposed tissue, or reluctance to scratch and eat. This is a more serious injury and does require some help from you.

TREATMENT

1. Provide a safe space for the chicken to heal away from the flock. Offer softened food and fresh, clean water.
2. Gently rinse the area with saline or a mild poultry-safe antiseptic to remove dirt and debris. Do not let the solution pour into the nostrils or down the throat.
3. If the beak is cracked, it may need support during healing. As long as there isn't active bleeding or exposed tissue, you can very carefully use a thin layer of gauze or piece of a tea bag to secure the area. Disinfect the area, then use a nontoxic superglue to attach it to the outside of the beak. It is very important that you have a helper hold the chicken securely and that you avoid getting any glue into the mouth, eyes, or nostrils. A vet could assist if you are not comfortable.
4. Monitor the chicken for signs of infection, starvation, and depression. Seek medical help if needed. Only integrate the chicken back into the flock after they are acting and eating normally.

Severe Break or Injury

If your chicken's beak is at an irregular angle, bleeding, has exposed tissue or bone, has a deep crack, or is partially hanging, you have a serious injury on your hands. A beak injury that reaches the inner layers of tissue is very painful and must be addressed quickly.

TREATMENT

1. Remove the chicken from the flock to avoid pecking and further injury. Plan to set up a space for your chicken where they can stay overnight for a few days to a few weeks.
2. Clean the area with a poultry-safe antiseptic spray to remove debris and help prevent infection.
3. Veterinary care is recommended for severe injuries so the chicken can be provided with antibiotics, pain relief, and possible reconstruction or protection of the soft tissue. In some severe cases, subcutaneous fluids may even be required to prevent dehydration. If healing isn't possible, the vet will be able to recommend humane euthanasia.
4. Keep the chicken separate from the flock until the risk of reinjury has passed and the bird is eating and drinking properly.

Prevention

- Eliminate hazardous areas that could cause snagging, such as broken wire mesh, protruding nails and screws, or sharp edges.
- Provide large rocks or cement slabs for natural filing of beaks. Trim beaks that are overgrown with a rotary filing tool or pet nail trimmers. Carefully remove small sections at a time to avoid cutting into the tissue.
- Provide ample space for chickens to move, flap, jump, and run.

CROSS BEAK

CiCi was born with a skull and beak deformation that makes it impossible for her to eat without my help. Because she relies on me to eat, she is my most tame, friendly chicken.

I secure CiCi between my knees and feed her chicken torpedoes a few times a day!

CROSS BEAK, OR SCISSOR BEAK, is when the upper and lower parts of the beak do not align. It can range from a minor to very severe misalignment. Although a cross beak isn't necessarily an ailment, like a broken beak, I still wanted to provide information on how to handle the situation. Having a beak that doesn't line up properly can mean the chicken is unable to eat, drink, and preen. Chicken keepers who keep a cross-beak chicken will need to be prepared to assist with these tasks.

Causes

- Incorrect positioning of the embryo in the egg.
- Improper incubation conditions, such as humidity, temperature, or lack of turning.
- Nutritional deficiencies in growing chicks.
- Injury or trauma to the face or skull early in life.
- Some genetic lines or breeds have a higher likelihood of developing a cross beak.

Symptoms

- Visible misalignment of the beak, such as overlapping or crooked upper or lower beak. The alignment usually starts off small or unnoticeable and worsens over time.
- Difficulty eating and drinking, resulting in slow growth and low body weight.
- Increased spillage of feed or mess around the face.
- Failure to thrive or dehydration.
- Issues with preening, resulting in ragged feathers or pinfeathers that aren't fully unsheathed.

Treatment

There is no treatment to correct the beak misalignment. Many have suggested surgery to break and set the beak correctly, but this procedure would be extremely difficult to recover from. The cross beak is usually caused by a deformation of the skull, which would be a very in-depth surgery and unlikely to be successful. There are a few ways you can feed your bird. Monitor your chicken to be sure it is picking up food, gaining weight, and growing in proportion to the other chickens of the same age.

1. **Deep dish feeding:** Some cross-beak chickens can successfully pick up food if you provide a deep dish. This helps with scooping the feed.
2. **Wet mash feeding:** Provide wet mash or soaked feed to make it easier for the bird to eat. This requires patience, as your chicken may take a long time to get enough food, and you may need to clean their face feathers afterward.
3. **Chicken torpedo feeding:** This is my preferred method for my cross-beak chicken because it isn't messy, and I can weigh her food to ensure she is getting enough. I mix ground chicken feed, raw egg, and a little melted coconut oil to make a dough. I form the dough into torpedo shapes that are about 1 inch (2.5 cm) long. I place my chicken between my legs to secure her wings and body, then use one hand to hold her beak open and the other to place the torpedo into her mouth. I let her guide the torpedo all the way down so I don't push it into her trachea. It takes time and patience to learn the technique, but it can be very successful.
4. **Tube feeding:** Tube feeding requires precise control to prevent the chicken from aspirating. A chicken has an esophagus and a trachea opening in their mouth. Forcing food into the trachea will result in aspiration and likely cause death. It is best to reach out to your trusted veterinarian or experienced mentor to learn the procedure firsthand. Tube feeding tends to be more time-consuming than other feeding methods because the feed is very watered down, and they need to be fed several times a day to get an adequate amount.

Supportive Care

1. Monitor your chicken to be sure they are picking up food, gaining weight, and growing in proportion to the other chickens of the same age. Weigh your chicken weekly during growth periods.
2. Provide open or cup waterers, as your chicken may not be able to use nipple waterers.
3. Provide feathers and hygiene support. Preening may be difficult, so check feathers often and remove stuck-on dirt, feed, and droppings. Help during molting season by gently rolling grown-out pinfeathers between your fingers to remove the waxy sheaths.
4. Gently trim the excess beak growth. Be careful not to trim too far, or you might hit the blood supply. Trimming the beak will not usually help with eating, but it may prevent breakage.

Prevention

- Avoid breeding birds with known cross beak or cross-beak heredity.
- Control incubation temperature and humidity, and use secondary monitoring tools in case one fails.
- Provide balanced nutrition and give chicks plenty of space in the brooder to avoid bullying while they are growing.

CONCLUSION

AS YOU HAVE SEEN in this chapter, external ailments like lice, mites, injuries, wounds, and bumblefoot pose a very real threat to your flock. Being prepared with a good coop design, a well-stocked first aid kit, vigilance, and knowledge will help you navigate these concerns. That attentiveness is part of what it means to keep chickens responsibly and humanely.

But even a healthy-looking chicken on the outside may harbor hidden challenges or internal ailments. Because the reproductive and internal systems are so central to a hen's life and productivity, it's vital to understand how viruses, parasites, nutrition, and internal stressors affect whether the bird can lay well, fight off illness, and live comfortably.

In the next chapter, we'll shift focus inward to explore common internal ailments, reproductive issues, internal parasites, and hidden illness. With an understanding of external and internal well-being, you'll be better equipped to care for your flock and maintain good welfare.

The inner workings of a chicken are more complicated than you might expect.

6

COMMON INTERNAL *Ailments*

In this chapter, we shift focus from external wounds, parasites, and injuries and turn to internal systems that keep your flock alive and productive. You'll learn about common internal challenges, including parasites, nutritional imbalances, crop issues, toxicities, and cancer. Because internal ailments can strike quietly and escalate without clear outward signs, keeping a close eye on your flock is one of your most powerful tools.

As you begin this chapter on internal ailments, remember this vital point about chickens: They are masters at hiding illness. Remember, because chickens are prey animals, even minor discomfort or disease is often masked instinctively. Pay attention to subtle changes, as they may be your first clue to a larger problem. With knowledge, vigilance, and good husbandry practices, you can catch many problems early. This will allow you to intervene effectively and maintain the long-term health and productivity of your flock.

DECODING DROPPINGS: WHAT POOP REVEALS

CHICKENS OFTEN MAKE their most honest "reports" in their droppings. For a flock keeper, observing and interpreting what comes out can be one of the most powerful early warning signs of internal trouble. Chickens can hide their symptoms really well, but the one thing they can't hide is their poop.

A normal chicken dropping

Normal Dropping

A healthy, normal drop will be tan, brown, or grayish. It will be firm to soft with a distinct form and have a whitish cap or stripe, which is the urate. It will become looser than usual if the chicken has treats with a high-water content, but should return to normal consistency once the food has been digested.

A cecal dropping. Totally normal but less frequent than regular droppings.

Cecal Dropping

Cecal droppings are also a normal part of digestion, but appear less frequently. You will see cecal droppings several times a day. They are varying shades of brown and usually have a strong odor. The consistency is a bit sticky and looks a bit like chocolate pudding (I probably just ruined pudding for you; sorry). While it's normal to see every eight to ten droppings, it would be concerning if you were seeing them every dropping and your chicken was lethargic or not eating.

One of my followers sent this photo of a dropping after the flock had blueberries.

Colored Dropping

If the flock consumes more greens, berries, or other dark-pigmented food, their droppings can change color. It can be quite alarming to find a dark purple or bright green dropping in the coop, but think back to the last day or so to determine whether it could be from a treat they were given or something they foraged. My flock always finds honeysuckle berries, which make their droppings reddish.

A foamy dropping

Loose, Foamy, or Yellow Dropping

While this could be a totally normal dropping, especially after sudden diet changes or rich treats, it could be concerning if it doesn't return to normal quickly and could indicate a digestive issue or infection, such as coccidiosis or a parasite. Pay attention to your chicken's behavior and appetite in this instance and consult a vet if the chicken is declining fast. You can also try treating for coccidiosis (see page 108) if symptoms match.

White watery dropping

White or Watery Urates

Occasionally, a chicken will expel a dropping that looks like mostly water and urates (the whites of the poop). This can be normal, especially after a lot of water on a hot day or after treats such as watermelon. If those things aren't happening and you see it appear over and over, it could be caused by parasites, infections, kidney stress, dehydration, or other systemic issues.

Intestinal shedding

Intestinal Shedding

Intestinal shedding is the natural shedding of the intestinal lining. It happens as epithelial cells are shed and replaced. Red or pink-tinged mucus in a dropping can be normal once in a while, but if you see it frequently or in heavy amounts, it can indicate a problem such as heavy inflammation, infection, irritation, or parasites. Intestinal shedding usually has some shape to it, while blood from a larger issue looks like a pool or a blood mixture in the dropping.

Bright green dropping

Bright Green Dropping

Bright green droppings unrelated to food intake may indicate digestive upset, liver or gallbladder stress, or more serious gut issues, such as viral, bacterial, or parasitic infections. If you see this, look for lethargy, reduced appetite, and dehydration. Check their diet and external stressors, and make sure their water is clean. You can have their droppings tested for parasites or treat preemptively. It is important to consult a vet if the problem persists.

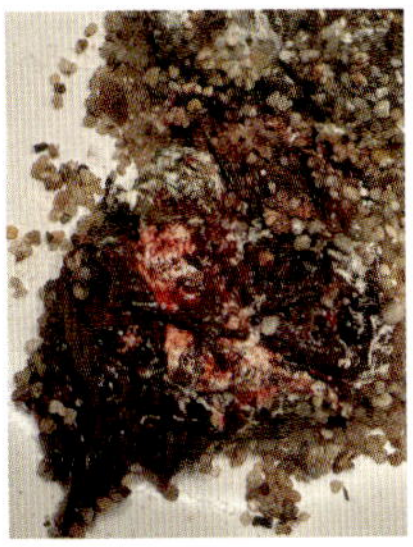
Bloody dropping

Bloody Dropping

The presence of bright red blood should be a warning sign for any chicken keeper. It is important to collect a sample and consult your vet if you are unsure of the cause, especially if your chicken is experiencing other symptoms. Blood in the droppings could indicate internal bleeding from ulcerations or reproductive stress. It could also be from parasites or severe intestinal disease, such as coccidiosis.

Black dropping

Black or Tarlike Droppings

Black droppings are cause for concern, as they can indicate bleeding in the upper digestive tract, severe gut damage, exposure to toxins, or contaminated feed. Act quickly to isolate the chicken, check for trauma or poisoning, inspect feed for mold or other toxins, and consult a veterinarian.

Worms

Mucus, Undigested Feed, or Worms/Worm Segments

Every chicken keeper's nightmare is worms in the poop! If you are seeing consistent droppings with mucus, undigested feed, diarrhea, weight loss, decreased egg production, poor appetite, gasping or neck stretching, and dirty vents, your chicken may have worms.

INTERNAL PARASITES—WORMS!

Healthy chicken, free from parasites

EVEN IN A WELL-KEPT FLOCK, worms are one of the most common and consequential parasites your chickens will face. Infestations are rarely obvious because they lurk quietly inside the chicken. They steadily steal nutrition, irritate tissues, and chip away at the bird's health. Only as the infestation grows will you possibly notice the effects on the birds or the worms themselves. Worms can go undetected for a long time, so proactive, preventive treatments will help maintain a healthy flock.

BE WARY The internet is full of home "remedies" for preventing and treating worms. It is important to note that ingredients such as red pepper flakes, pumpkin seeds, garlic, apple cider vinegar, oregano, and others lack scientific evidence supporting their effectiveness in preventing or treating worms. Often, when people say, "I have used this for years, and my chickens have never had worms," it's because worms generally aren't visible until an infestation is severe. I haven't used any of these products, and my chickens haven't had any visible worms either.

Roundworms

A follower sent me this picture. Notice the tapeworm segments in the dropping.

If your chicken is constantly gasping, it would be a good idea to check for gapeworm.

Types of Worms

ROUNDWORMS

Roundworms in poultry usually refer to nematode parasites, *Ascaridia galli* and related species, which are found in the small intestine of chickens. These worms can go unnoticed in low-level infections. Heavy infestations have more noticeable symptoms and can impair digestion, growth, and nutritional absorption. A heavy infestation may also be indicated by large roundworms in the droppings. Roundworms live and reproduce inside the bird. They shed their eggs, which end up in the droppings, spreading the infection into the environment and to other birds. Left untreated, a chicken can experience severe symptoms and even death.

CECAL WORMS

Cecal worms live in the ceca of chickens. Ceca are two pouches located near the small and large intestines that house bacteria for breaking down food. While mild cecal worm infections don't always show obvious signs, heavy infestations can be more problematic. Cecal worms can also carry the pathogen that causes blackhead disease. Cecal worms are not always detectable with the naked eye.

HAIRWORMS/THREADWORMS

Capillaria refers to a group of very thin nematode (roundworm) parasites. They are often called hairworms or threadworms, and they infect different parts of the chicken's digestive tract. This could be the crop, esophagus, small intestine, ceca, or lower gut. They are small and often overlooked, but they can pose a serious threat to your flock. The chicken could experience diarrhea, or droppings may contain blood or mucus.

TAPEWORMS

There are many species that can infect poultry, but the most common are Raillietina and Davainea. Tapeworms live in the intestines of the chicken by attaching themselves to the intestinal wall. Tapeworms are composed of many segments, each capable of producing eggs. The segments are shed, and you may see them in the droppings. Free-range chickens, or those with access to soil and insects, are at higher risk of contamination. This is because tapeworms need an intermediate host to complete their life cycle. Beetles, earthworms, flies, slugs, and other insects ingest the eggs and then get eaten by the chicken. Tapeworms compete with chickens for nutrients, but chickens can tolerate a light infestation without showing any signs. A heavy infestation could cause inflammation, intestinal lesions, and extreme health decline. A clear sign of tapeworms is droppings with small segments present that look like small, pale grains of rice.

GAPEWORMS

Unlike most internal parasites that live in the digestive system, gapeworms lodge in the bird's respiratory tract. They can be found in the windpipe, trachea, or even deeper into the lungs. They feed on blood, which causes inflammation and mucus buildup. This leads to airway obstruction and makes the chicken gasp for air or open-mouth breathe. They will often stretch out their necks as they gasp. These infestations can lead to loss of appetite, suffocation, and death. Gapeworms are usually not visible when the beak is opened. Gasping can be caused by respiratory illness or adjusting to move feed through the system. If those have been ruled out, it may be best to treat for gapeworms.

General Symptoms of Worm Infestation

- Poor weight gain, slow growth, and poor body condition.
- Reduced appetite.
- Eating normally, but still losing weight.
- Pale combs, wattles, or face, indicating anemia.
- Diarrhea or droppings stuck around the vent feathers.
- Droppings that contain blood, mucus, or even visible worms or eggs.
- Lethargy, low activity, drop in egg production, or droopy wings.
- Intestinal lining damage, intestinal blockage, intestinal hemorrhaging, and even death.
- Symptoms are more pronounced in younger birds.

Diagnosis

Waiting for visible signs, such as dirty vents or lethargic birds, may not be the most effective way to monitor for worms. A chicken can carry a light worm load and show no obvious signs. It is best to use regular fecal testing. Fecal float testing can be done at home with a microscope or by a veterinarian. There are also companies that let you send a sample by mail. A fecal test will help you determine whether or not deworming is needed. Sometimes you can see the eggs just by looking, and in that case, you can go ahead and deworm. Fecal float tests may produce a negative result, but worms may still be present. If symptoms persist, it is best to deworm your flock.

Treatment

1. Treat the whole flock with a product specifically for the type of parasite your flock is dealing with. Follow the label instructions for dosage.
2. The environment should also be thoroughly cleaned to prevent ongoing spread. Remove and replace bedding, clean the run substrate, and disinfect the feeders and waterers.
3. Repeat treatment if needed or as directed on the medication label.

DEWORMING CHART

PARASITE	MEDICATION	COMMON BRAND NAMES	TREATMENT TIPS
Roundworms	Fenbendazole	Safe-guard AquaSol	Treat the entire flock. No egg withdrawal if used as directed.
Cecal Worms	Fenbendazole	Safe-guard AquaSol	Treat the entire flock. No egg withdrawal if used as directed.
Hairworms/ Threadworms	Fenbendazole	Safe-guard AquaSol	Often requires higher dosing; 14-day egg withdrawal suggested.
Tapeworms	Praziquantel	Droncit	Off-label medication requires egg withdrawal; 14 days suggested.
Gapeworms	Fenbendazole	Safe-guard AquaSol	Higher dosing often required; 14-day egg withdrawal suggested.

Prevention

- Maintain a clean environment to prevent the spread among the flock. Don't allow feces to accumulate.
- Raise feeders and waters off the ground to prevent contamination from feces. Avoid feeding the chickens on the ground, especially in muddy conditions or when the ground is heavily soiled.
- Rotate free-range pastures when possible, or use a tractor-style coop to provide fresh foraging space.
- Quarantine and test new flock members, especially older ones that have been exposed to the outside.

ASK THE POULTRY DOC

DEWORMING

Waiting to see worms in droppings is unreliable and often too late. Many important poultry parasites, such as Capillaria and cecal worms, are microscopic or rarely visible. By the time adult worms are observed, birds may already be experiencing weight loss, diarrhea, or intestinal damage caused by earlier life stages.

The best method for parasite control is fecal testing with treatment guided by results. When diagnostics aren't available, a targeted course of a labeled dewormer at the correct dose is a reasonable alternative—not routine monthly deworming, but a risk-based intervention. Because most poultry roundworm life cycles are roughly 3 to 4 weeks, strategic prophylactic deworming about every 3 months may be reasonable in the higher-risk environments of backyard poultry keepers.

INTERNAL PARASITES: PROTOZOAL

MANY INTERNAL PARASITES are microscopic, single-celled organisms, not visible worms. These are known as protozoa and can be just as, if not more, dangerous to a flock. They act silently until severe damage occurs, multiplying and damaging tissues or interfering with nutrient absorption. Because they are invisible to the naked eye, it is essential to be aware of possible symptoms, monitor droppings, and manage hygiene.

Many of these protozoa are a regular part of the poultry environment. They are microscopic and live in the litter, soil, dust, etc. In many flocks, a low-level presence is inevitable and tolerable. Over time, they are exposed to small doses, allowing their immune systems to build resistance. The problem arises when poor hygiene, overcrowding, stress, or heavy exposure is in play.

CHICKS AT RISK! For some protozoal parasites, young chicks and growers are at a higher risk. Their immature immune systems are weak, making it harder to cope with the infestation. That being said, older birds are not immune. Adult chickens can still be infected, especially if they are living in crowded, dirty conditions. They may not show symptoms but can still be carriers, transferring it to young chickens when they join the flock.

COCCIDIOSIS

This hen is showing signs similar to coccidiosis. Notice her ruffled feathers, droopy posture, and pale skin.

COCCIDIOSIS IS A COMMON and often serious intestinal disease in chickens. The condition is caused by coccidia, protozoa that reproduce and spread via egglike oocysts. Oocysts are shed in droppings from infected birds and contaminate the environment. They are then ingested, infecting other birds. The oocysts burst open, invade intestinal cells, reproduce, and shed new oocysts in droppings. When these parasites invade the gut wall and rapidly reproduce, they damage the intestinal lining and interfere with nutrient absorption. It is one of the most widespread internal diseases of poultry worldwide. While most healthy flocks carry a small load of coccidia, coccidiosis occurs when the load becomes too heavy.

Causes

- Unkempt brooders.
- Exposing young chicks to the outside environment too soon.
- Overcrowding or stress.
- Damp, dirty, or soiled litter.
- Contaminated waterers or feeders.
- Poor hygiene.

Symptoms

- High incidence of diarrhea, sometimes with mucus and, in severe cases, blood.
- Decreased appetite, lethargy, ruffled feathers, and droopy posture.
- One or more chickens huddled alone with their eyes closed.
- Pale combs and wattles.
- Reduced egg production.

Diagnosis

To confirm suspicions of coccidiosis, a fecal exam must be performed to detect oocysts. If the infestation is severe and birds are already showing major signs, it would likely be too late to send a sample by mail. Reaching out to your vet for an immediate appointment would be your best chance for quick detection. If that is not available, treating for coccidiosis would be warranted. If a chicken dies, a necropsy can determine whether it was due to coccidiosis.

Treatment

1. Anticoccidial drugs may be administered to the affected birds or the whole flock. Amprolium is the most commonly used in backyard flocks. It is administered in the drinking water for 3 to 7 days.
2. Support the flock after treatment with a poultry vitamin mix in their water and nutritional chicken feed.
3. Clean the environment of soiled bedding, oversaturated litter, and dirty waterers.

Prevention

- Vaccination is available for chicks and can help build immunity.
- Newly hatched chicks that aren't vaccinated can be fed medicated feed for a few weeks to help stabilize their systems as they slowly get introduced to the parasite.
- Watch young birds closely as they are introduced to the outdoors and new coops. Keep their environment clean and don't overcrowd.
- Keep bedding clean and dry. Put a roof on the chicken run to prevent muddy conditions where oocysts could thrive.
- Rotate outdoor foraging areas to break the coccidia life cycle.

ASK THE POULTRY DOC

DOSAGE FOR AMPROLIUM

When managing coccidiosis, diagnostic confirmation can be extremely helpful in maintaining long-term flock health. Before initiating any treatment, it is always best to rely on the results of a fecal flotation test to quantify the presence of coccidia. Because a low level of oocysts is a normal finding in most healthy birds, the decision to treat should be based on a combination of clinical signs and the specific parasite load. It is important to note that the threshold for treatment varies; some species of Eimeria are significantly more fecund (producing a high volume of eggs) without causing major harm, while others may produce fewer eggs but result in much more severe intestinal lesions.

For birds requiring intervention, the recommended dosage for amprolium is 2 teaspoons (10 ml) per 1 gallon (3.8 L) of drinking water, supplied as the sole source of hydration for 3 consecutive days. Providing a higher dose or extending the duration of treatment can be counterproductive, as it may interrupt the bird's ability to develop natural, lifelong immunity, potentially leading to more severe outbreaks later in life.

HISTOMONIASIS: BLACKHEAD DISEASE

HISTOMONIASIS, MORE COMMONLY known as blackhead disease, is another protozoal disease caused by *Histomonas meleagridis*. It isn't as prominent in chickens as it is in turkeys and other game birds, but chickens can still suffer from the disease and, at the very least, be carriers. It gets its name from the rare occurrence of the bird's head and comb turning blue or black from poor oxygenation.

Causes

- Birds ingest eggs of the cecal nematode *Heterakis gallinarum* (common cecal worm). These eggs may introduce the protozoa to the body. Eggs can be found in droppings, contaminated soil, feed, or litter.
- Transport hosts, such as earthworms, insects, or soil invertebrates, may carry it, making free-range birds susceptible.

Symptoms

- Lethargy, depression, ruffled or dull feathers, droopy posture, and general unwellness.
- Reduced appetite, poor growth, or failure to thrive.
- Yellowish diarrhea.
- In more severe cases, you may see long-term underperformance.
- Turkeys or game birds may experience high mortality and the signature blackening of the head and face.

Diagnosis

Diagnosis usually requires fecal analysis or tissue sample analysis in a lab or under a microscope. Contact a veterinarian if chickens show signs of blackhead disease. If a chicken has died from blackhead, a necropsy will usually show damaged or necrotic tissue in the ceca or target-shaped lesions on the liver.

Treatment

Unfortunately, there is no approved or effective treatment for histomoniasis. Many drugs that were previously used are now restricted. The best course of action if your flock has confirmed blackhead disease is to prevent the spread (especially to turkeys or game birds) and manage their environment and sanitation.

Prevention

- Maintain a clean environment to prevent the spread among the flock.
- Quarantine and test new flock members, especially older ones that have been exposed to the outside.
- Avoid overcrowding.

EAR INFECTIONS Did you know chickens can get ear infections from viruses, bacteria, and fungi? Ear infections can cause swollen or crusty ear openings, head shaking, head tilt, ear scratching, and balance issues. Prompt veterinary care, cleaning, and targeted antibiotics help resolve the infection.

TRICHOMONIASIS: CANKER

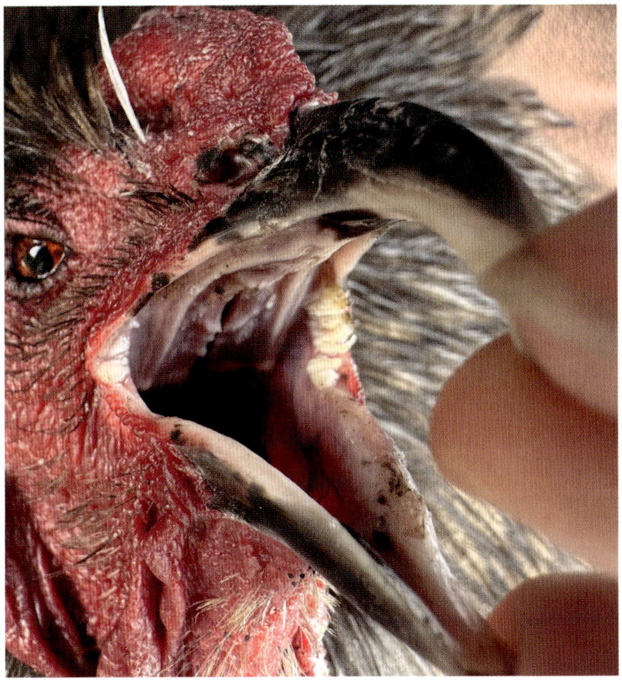

Trichomoniasis canker

TRICHOMONIASIS IS CAUSED BY the protozoan parasite called *Trichomonas gallinae*. It can affect pigeons, doves, game birds, chickens, turkeys, and other poultry. Many people refer to it as "canker." It tends to take hold and develop in the upper digestive tract rather than the intestines. It shows up as lesions or a cheeselike buildup in the mouth and throat. Canker causes difficulty breathing, swallowing, and eating.

Causes

- Direct contact with contaminated shared feeders and waterers.
- Close quarters, overcrowded pens, or multiple species being kept together.

Symptoms

- Soft, cheesy-looking, or yellowish caseous mass or lesion in the mouth or further down in the throat.
- Difficulty eating, drinking, or swallowing.
- Reduced feed intake, causing weight loss.
- General poor condition, decreased activity, and listlessness.

Diagnosis

To find canker, the chicken keeper must inspect the mouth and throat for lesions or yellow masses. It is easier with a helper to secure the bird and pry open the beak. Use a flashlight to look into the mouth. A vet can take a mouth swab to examine it under a microscope.

Treatment

1. Separate the infected bird, disinfect shared feeders and waterers, and change bedding.
2. Call your vet for an examination, as the mass may need to be removed. The vet may prescribe medications such as Carnidazole, Metronidazole, or Toltrazuril, depending on your location, regulations, and availability. There is some literature to support the use of thyme extract.

Prevention

- Keep chicken waterers and feeders away from wild birds, especially pigeons and doves.
- Maintain a clean coop environment with clean waterers and feeders.

COMMON INFECTIOUS DISEASES: BACTERIAL

THE MOST COMMON BACTERIAL infections in poultry are salmonellosis, coryza, cholera, and mycoplasmosis.

SALMONELLOSIS

SALMONELLOSIS is a bacterial infection caused by Salmonella. Salmonella is a serious pathogen in poultry flocks. There are various strains; some are associated with poultry, while others may infect other species. When a chicken becomes infected, they may not show symptoms but can still shed Salmonella in their droppings, making it a hidden threat. It can also be present in eggs and meat, posing a health risk to humans.

Causes

- Ingesting the bacteria found in contaminated feed, water, bedding, soil, droppings, or equipment.
- Mother hens pass it to their offspring through the reproductive tract.
- New birds are integrated without quarantine or testing.

Symptoms

- Loose droppings, diarrhea, and possible dirty vents.
- Failure to thrive, slow growth, and poor feed conversion.
- Lethargy, weakness, and a general appearance of unwellness.
- Hatch-rate issues that are causing decreased hatchability or reduced egg production.
- Poorly developed chicks.
- Systemic illness and sudden death.

Diagnosis

Fecal, cloacal, or tissue samples can be sent to a lab for bacterial culture to detect Salmonella. If the chicken has died, specific organs may reveal lesions or signs of infection. Eggs or chicks may also be tested to determine whether a hen has an infection in the reproductive tract.

Treatment

1. Antibiotics may be used to treat Salmonella, but many strains are resistant to them. There is a withdrawal time before eggs and meat can be consumed, and birds can be reinfected if they remain in the environment.
2. Keep the infected bird isolated during treatment. Ensure good nutrition and fresh water. Focus on gut health with prebiotics and probiotics.
3. Clean and disinfect the coop, litter, feeders, and waterers with a poultry-safe disinfectant. Move the coop and run to a new location if feasible.

Prevention

- Rotate pastures for free-range or pasture-fed birds.
- Use a strict regimen for cleaning the coop, run, feeders, and waterers.
- Practice good biosecurity by quarantining new chickens and not introducing pathogens on your own clothes or boots that have visited other farms.
- Salmonella grows faster in warmer temperatures, so store eggs in the refrigerator after collection. Once they are cold, they must stay cold until use. Avoid eating cracked eggs. Wash eggs before use and cook them thoroughly.

INFECTIOUS CORYZA

Chicken with coryza

INFECTIOUS CORYZA, commonly called "coryza" for short, is an acute, highly contagious bacterial disease. It affects a chicken's upper respiratory tract and can affect chickens of all ages. Coryza spreads quickly and becomes difficult to manage as recovered birds can still carry and shed the bacterium. Even if they appear healthy, previously infected chickens can still be carriers. The incubation period is short, and the onset of disease is rapid.

Causes

- Direct contact with an infected bird, respiratory droplets, or a contaminated environment.
- Weakened immune system.

Symptoms

- Coughing, sneezing, watery eyes.
- Nasal discharge.
- Swelling of the intra-orbital sinuses.
- Puffy, swollen face.
- Respiratory distress.
- Reduced feed intake.
- Lethargy.
- Drop in egg production.

Diagnosis

A tentative diagnosis begins with symptoms such as facial swelling, nasal discharge, ocular discharge, and respiratory signs that quickly spread to other birds. A confirmed diagnosis needs a bacterial culture to identify the bacterium.

Treatment

1. Coryza is a bacterial infection, so antibiotics are typically effective. Follow the vet's guidance for administering the medication.
2. Support the flock by ensuring clean drinking water, giving nutritional feed, improving coop hygiene, and reducing stress.
3. Antibiotics suppress symptoms but do not guarantee the elimination of the bacterium from all birds. Continue with improved coop management to ensure a clean environment.

Prevention

- Perform adequate biosecurity measures. Isolate new chickens, disinfect footwear and equipment when moving between pens, and avoid open waterers and use nipple waterers instead.
- Clean and disinfect coops, runs, feeders, and waterers. Maintain a clean coop by removing soiled bedding and ensuring good ventilation.
- Minimize overcrowding.

FOWL CHOLERA

FOWL CHOLERA IS A CONTAGIOUS bacterial disease of poultry and other bird species caused by *Pasteurella multocida*. It can be sudden with high mortality, a chronic localized infection, or the bird can be a carrier with no signs. If the infected chicken survives, it can still be a carrier and shed the bacteria. Cool, moist conditions allow the bacterium to survive in the environment for weeks. This makes it difficult to eradicate without proper biosecurity and sanitation.

Causes

- Direct contact with an infected bird.
- Transmission through mucus, saliva, and contaminated surfaces such as waterers and bedding.
- Indirect contact with contaminated equipment or clothing.
- Wild animals can act as carriers, transferring it to the flock.
- Overcrowding, poor sanitation, stress, damp weather, and the integration of new chickens without quarantine.

Symptoms

ACUTE FORM (SEPTICEMIA)

- Sudden death with little to no warning signs.
- Ruffled feathers, lethargy, and reluctance to move or eat.
- Labored breathing and nasal discharge.
- Diarrhea that changes to green over time.
- Darkening of the comb or wattles to a purple or bluish color.

CHRONIC, LOCALIZED FORM

- Lameness or difficulty walking due to joint problems.
- Respiratory signs if the upper airways are involved.
- Swelling or abscesses in areas such as the head, face, wattles, joints, footpads, or ears.

Diagnosis

Live birds can be tested by a veterinarian with a swab of the sinuses, oral cavities, or abscesses. Birds that die suddenly or show symptoms such as abscesses or other signs of fowl cholera can be diagnosed by testing affected tissues. In some cases, veterinarians will examine gangrenous tissues or organs to confirm the infection.

Treatment

1. Fowl cholera is bacterial and therefore can be treated with antibiotics prescribed by a veterinarian.
2. Treatment does not guarantee elimination, as survivors can become carriers, leading the flock to fall sick again.
3. For long-term control, prevention through good sanitation and strict biosecurity is recommended.

Prevention

- Rodent, wild birds, and predator control are necessary with a secure coop and run.
- Quarantine new birds before bringing them into your flock.
- Clean and disinfect coops, feeders, waterers, and bedding after an outbreak.
- Maintain good coop hygiene and avoid stressors such as overcrowding.

MYCOPLASMOSIS

When my chicken had a swollen eye, I started to worry she had mycoplasmosis, but it was just a simple respiratory infection.

MYCOPLASMOSIS IS A BACTERIAL DISEASE caused by *Mycoplasma gallisepticum* (MG) or *Mycoplasma synoviae* (MS). Chickens infected with MG develop upper respiratory symptoms and often get chronic respiratory disease. Chickens with MS usually have mild respiratory symptoms but may also develop joint, tendon, leg, and egg-production issues. Mycoplasma infections are often chronic and may persist for life. Even healthy birds can be carriers and spread the infection. Mycoplasma infections can weaken the immune system and lead to secondary infections, making symptoms more severe and persistent. Mycoplasmosis has a low mortality rate unless compounded with other pathogens.

Causes

- Mycoplasma bacteria entering the system and colonizing in the respiratory tract.
- Direct contact with infected birds or carriers.
- Contaminated equipment or bedding, shared feeders and waterers.
- Parent-to-chick transmission via egg.
- Stressors such as overcrowding and poor hygiene.
- The addition of new flock members with mycoplasmosis.

Symptoms

MG AND MS

- Sneezing, coughing, wheezing, or gurgling often with open-mouth breathing.
- Nasal discharge, watery or foamy eyes, conjunctivitis, or eye discharge.
- Mild or occasional swelling around the head, sinuses, and eyes.
- Decreased egg production and poor hatchability.
- Lethargy, reduced eating and drinking, ruffled feathers.

MG

- More pronounced respiratory distress, sinus and airway inflammation, and chronic respiratory disease.

MS

- Joint problems, tendon inflammation, footpad issues, and lameness.
- Egg production problems or shell abnormalities.

Diagnosis

MG and MS resemble other respiratory illnesses. Proper treatment requires testing. The common test is a real-time polymerase chain reaction (PCR) using swabs from the trachea, oropharynx, or sinuses. You can order tests online or visit a vet.

Treatment

1. Most antibiotics target cell walls, but mycoplasmas lack one, making many antibiotics ineffective. An appropriate antibiotic is needed, so testing is advised. While antibiotics reduce symptoms, they don't guarantee bacteria elimination, and birds remain lifelong carriers.
2. Maintain strict biosecurity, maintain good hygiene, ensure ventilation and dust control. Minimize stress.

Prevention

- Quarantine new birds.
- Maintain proper cleaning and hygiene.

COMMON INFECTIOUS DISEASES: VIRAL

THE MORE COMMON VIRAL infections include Marek's disease, Newcastle disease, infectious bronchitis, fowl pox, and bird flu.

MAREK'S DISEASE

MAREK'S DISEASE, caused by a herpesvirus, is widespread in chickens worldwide. It spreads easily through dust and feather dander, so most flocks not in sterile conditions are presumed exposed. Once infected, chickens remain so for life, shedding the virus regardless of symptoms. Egg layers are more affected than meat birds, and Silkies are highly susceptible. Symptoms usually appear between 2 and 6 months, but older chickens can also be affected. Marek's affects nerves, immune cells, and organs, with symptoms varying by strain and tissue involvement.

Causes

- Inhalation of virus-laden dust or dander from other chickens.

Symptoms

- Leg or wing paralysis or lameness that is oftentimes asymmetric, causing the bird to be unsteady.
- Weight loss or stunted growth.
- Tumors or enlarged nerves and internal organs visible during a necropsy.
- Skin or feather follicle lesions.
- Ocular changes are possible, as well as immunosuppression.

Diagnosis

Clinical signs, along with the flock history, are two factors that could support a temporary diagnosis. Flock keepers can mail feathers to a lab for a PCR test. A necropsy could diagnose Marek's as well by looking for characteristic tumors and nerve lesions.

Treatment

Young chicks can be vaccinated, but there is no cure for Marek's disease once clinical signs appear. Supportive care may keep the chicken comfortable, but recovery from nerve damage or tumors is improbable. Supportive care may include proper nutrition, isolation, soft bedding, warmth, and access to water. Chickens may deteriorate quickly, in which case humane euthanasia should be considered.

Prevention

- Day-old chicks can be vaccinated to ensure it is done before exposure.
- Strict biosecurity could keep your flock safe by preventing the introduction of unvaccinated birds.

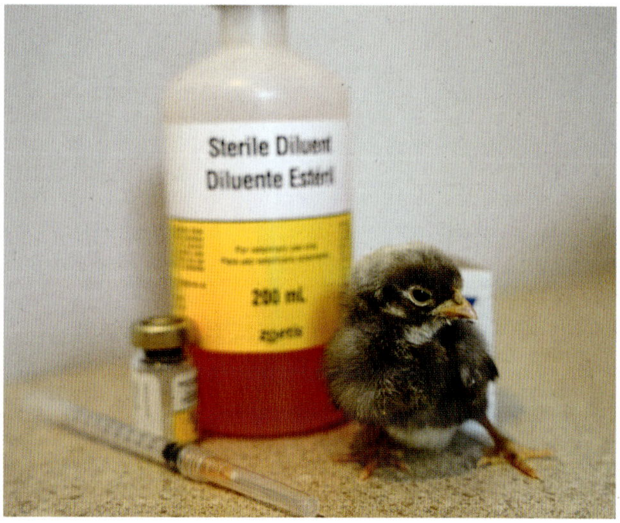

Vaccination of a chick

NEWCASTLE DISEASE

NEWCASTLE DISEASE is a highly contagious viral disease that affects the respiratory, nervous, and digestive systems. Severity ranges from mild illness to sudden death, especially in unvaccinated flocks. The virus can live for several weeks in cool environments or bedding, but it does not survive long in sunlight.

Causes

- Contact with infected respiratory secretions, saliva, and feces, including from wild birds.
- Contaminated equipment, footwear, or clothing.

Symptoms

- Coughing, sneezing, labored breathing, or nasal discharge.
- Trembling, twisted neck, paralysis, or drooping wings.
- Green droppings or diarrhea with blood.
- Low egg production, thin shells, or misshapen eggs.

Diagnosis

A confirmed diagnosis requires lab testing by a veterinarian or state official using swabs from live or deceased birds.

Treatment

1. Unfortunately, there is no medication to eliminate Newcastle disease after infection. Isolate infected birds, provide supportive care, and clean and sanitize the coop and chicken yard.
2. Some places require the outbreak to be reported to local public health officials.

Prevention

- Vaccination is widely used in commercial poultry operations, and in some areas, it is available for backyard flocks.
- Biosecurity is important when moving between farms or handling birds from unknown backgrounds. Isolate new chickens.

INFECTIOUS BRONCHITIS

INFECTIOUS BRONCHITIS is a highly contagious viral respiratory disease affecting chickens worldwide. It spreads rapidly through flocks and can be especially severe in young birds, sometimes leading to high mortality.

Causes

- Contact with airborne droplets or infected surfaces.
- Contaminated clothing or equipment.

Symptoms

- Coughing, sneezing, and nasal and ocular discharge.
- Difficulty breathing or gasping.
- Lower egg production, thin shells, or misshapen eggs.
- Ruffled feathers or lethargy.
- Poor growth, low appetite, or dehydration.

Diagnosis

Because many respiratory illnesses share or overlap symptoms, lab testing is required to confirm the diagnosis. Chickens with symptoms that are spreading quickly through the flock should be tested for infectious bronchitis.

Treatment

1. There is no cure for the virus, so provide supportive care such as warmth and electrolytes if needed.
2. Secondary infections may arise due to a weakened immune system, so medication or care may be needed.

Prevention

- Vaccination is available for commercial flocks.
- Maintain proper biosecurity and limit exposure to wild birds. Don't share equipment between flocks or farms.
- Keep the coop clean and provide your flock with healthy nutrition.

FOWL POX

Fowl pox can be very uncomfortable for chickens, as evidenced in this image from one of my followers.

FOWL POX IS A COMMON VIRAL DISEASE in backyard flocks, caused by various strains of DNA poxvirus. It has two main forms: dry pox, which affects the skin, wattles, and combs, producing wartlike bumps; and wet pox, which affects the mucous membranes of the mouth and throat. Most chickens survive, develop immunity, and are not carriers after infection. The virus can persist in the environment for a long time, especially when scabs or lesions remain.

Causes

- Direct contact with other infected birds.
- Mosquito and fly bites.
- Contaminated surfaces such as feeders, coop dust, or equipment.

Symptoms

DRY POX

- Wartlike lesions or scabs on any exposed skin that appear as pale bumps but darken and crust over.
- Reduced food intake, ruffled feathers, a drop in egg production, and a general air of unwellness.

WET POX

- Yellowish, cheesy lesions inside the mouth, throat, windpipe, and sinuses.
- Difficulty eating and breathing.
- Some eye discharge.
- Reduced food intake, ruffled feathers, a drop in egg production, and a general air of unwellness.

Diagnosis

Dry pox is easy to spot, with its characteristic wartlike scabs on the combs, wattles, face, and sometimes the beak and legs. Wet pox is often confused with respiratory illness or canker. Lab testing may be required to confirm wet pox.

Treatment

1. Since this is a virus, there is no antibiotic cure. Treatment is supportive care, isolation of infected birds, and coop disinfecting.
2. Provide soft food and easy access to water to encourage eating and drinking, especially if mouth lesions are making it difficult.
3. Veterinary care may be needed if difficulties arise or secondary infections take place.

Prevention

- A vaccine is available for backyard chicken keepers. The vaccine can be ordered online and administered at home with the wing stick method. It is recommended to administer the first dose when they are 6 to 12 weeks old. A booster is needed before they start laying eggs between 18 and 20 weeks. Some sources say a yearly booster is warranted in areas where fowl pox is common. Other sources say the two doses are enough to offer lifelong immunity.
- Control mosquito and fly populations by regularly eliminating standing water, food scraps, and droppings.

AVIAN INFLUENZA

Although this chicken didn't have avian influenza, many respiratory illnesses present with symptoms such as bubbly eyes.

AVIAN INFLUENZA (AI), OR BIRD FLU, is caused by the influenza A virus and can be either high or low pathogenic. High pathogenic strains kill flocks quickly, while low pathogenic strains cause mild respiratory issues. It infects wild birds, especially waterfowl, and can spread to poultry and other animals. Wild birds, especially during migration, pose a significant risk; proximity to water bodies increases the danger. Rarely, humans can be infected with mild or no symptoms. AI is a serious threat to backyard flocks.

Causes

- Transmission through direct contact with infected domestic or wild birds.
- Infected saliva, nasal secretions, or droppings.
- Shared water sources with infected chickens.
- Wild birds sharing water sources or flying overhead.

Symptoms

LOW PATHOGENIC AI

- Mild respiratory symptoms such as sneezing, coughing, and nasal discharge.
- May show no signs of illness.
- Ruffled feathers, lack of energy, and low appetite.
- Possible diarrhea.
- Decreased egg production, some soft-shelled eggs.

HIGH PATHOGENIC AI

- Possible sudden death and high mortality rate.
- Rapid spread among the flock and other flocks in the area.
- Severe respiratory distress, such as gasping, coughing, sneezing, and nasal discharge.
- Ruffled feathers.
- Lack of energy, coordination issues, and low appetite.
- Significant drop in egg production and an increase in soft-shelled eggs.
- Swelling and purple discoloration of the face, wattles, combs, and legs.
- Possible diarrhea.

Diagnosis

If your flock is experiencing respiratory signs, rapid transmission, and sudden death, you can suspect AI. To confirm the diagnosis, a veterinarian or animal authority would need to collect samples and send them to a lab for testing, especially because these signs overlap with those of other diseases. If high pathogen AI is confirmed, reporting to state health officials is likely required. After it is reported, authorities will take immediate action, including testing, quarantine, or euthanasia of affected flocks, to prevent the spread of the disease.

Treatment

1. There is no universal effective antiviral treatment for AI in backyard flocks. Because of its severity and transmission rate, the primary response is isolation, biosecurity, and culling of affected birds. These practices help stop the spread.
2. If you have sick birds, provide supportive care like clean housing, warmth, hydration, and good nutrition. Research shows oregano oil can boost immunity and reduce pathogens, aiding disease fight. However, chickens stay highly contagious.

Prevention

- Several countries like China, Egypt, Indonesia, and Vietnam use an avian flu vaccine. The United States and other countries prefer culling infected flocks, believing the vaccine prevents severe illness but not infection, which can still spread. Some worry vaccinated birds look healthy and spread the virus, complicating control. The United States is reevaluating its stance.
- Strict biosecurity is essential if your flock is in a waterfowl migration area. Limit contact with wild birds by covering the run and sealing small holes. Limit free-range access during high-risk seasons, secure feed and water, and prevent contamination from your own boots and equipment.
- Avoid introducing new birds from unknown backgrounds.

METABOLIC ISSUES

METABOLIC ISSUES to be aware of include nutritional deficiency, fatty liver hemorrhagic syndrome, and gout.

NUTRITIONAL DEFICIENCY

A chicken who eats a healthy diet also looks bright-eyed and alert.

THE PHRASE "NUTRITIONAL DEFICIENCY" refers to the problems that arise when a chicken's diet lacks sufficient essential nutrients. Calcium, phosphorus, vitamin D, magnesium, and protein are among the nutrients that play a vital role in a chicken's health. An imbalanced or inadequate diet can compromise egg production, bone health, immunity, growth, feather quality, and overall vitality.

Causes

- An imbalanced diet or one that is not age-appropriate.
- Feeding a diet that is homemade without proper poultry science to back its formula.
- Too many treats, kitchen scraps, scratch grain, or fruits and vegetables, causing chickens to eat less of their balanced chicken feed.
- Lack of access to clean water and good-quality feed.
- Free-range birds that are having trouble satisfying their nutritional needs from the land and are not fed chicken feed.
- Moldy or spoiled feed can make chickens refuse to eat or can affect absorption.

Symptoms

- **Vitamin A deficiency:** Poor vision, watery eye discharge, poor growth, weak feathers, decreased egg production, decreased hatchability.
- **Vitamin D3 deficiency:** Chicks may experience leg weakness, soft beaks and nails, and difficulty walking or standing. Adult chickens may deal with weak bones and poor eggshell quality.
- **Vitamin E:** Muscle weakness, muscular dystrophy, poor coordination, or paralysis.
- **B vitamin deficiencies:** Poor growth, poor feather quality, curled toes, loss of appetite, poor embryo health.
- **Calcium/phosphorus imbalance:** Weak bones or bone deformities, poor egg quality and soft shells, increased fractures or bone injuries.

ASK THE POULTRY DOC
VITAMIN DEFICIENCY

While vitamin deficiencies are frequently discussed in poultry circles, they are exceptionally rare in practice. Most descriptions found in textbooks or scientific literature regarding specific clinical signs are often the result of laboratory-controlled studies—environments where diets are intentionally formulated to be void of a single nutrient to isolate its physiological role. In the real world, a true vitamin deficiency is almost never an isolated event. Instead, it typically manifests as a broader symptom of malnutrition, appearing alongside poor feathering, stunted growth, and birds that are generally systemically unwell.

What many keepers interpret as a "vitamin deficiency" based on a single symptom is more often a neurologic or systemic disease with an underlying cause completely unrelated to nutrition. When birds are maintained on complete rations manufactured by reputable feed companies, they are receiving a diet scientifically formulated to meet their needs, including a healthy safety margin. Therefore, if a flock is being fed a proper, age-appropriate ration, keepers should look beyond the feed bag toward health and environmental factors when investigating clinical abnormalities.

- **Trace mineral deficiencies:** Poor eggshell quality, reduced hatchability, lower egg production, and general poor condition.
- **Electrolyte/salt imbalance:** Lower egg production, lethargy, poor growth, or general weakness.
- **Protein/amino acid imbalance:** Poor growth, stunted development, slow feathering, poor feather quality, weak muscles, reduced egg production, small eggs, poor hatchability, weakened immune system, and difficulty recovering from stress or sickness.

Diagnosis

Observation will be your best tool for determining whether your chickens lack balanced nutrition. Poor growth, weak bones, and poor egg quality are all good indicators that you should check their feed's nutritional content.

Treatment

1. Feed the chickens a balanced feed for their stage of life. Always feed the youngest, most fragile members of the flock.
2. Treats, kitchen scraps, and scratch feed should be a minimal part of their diet.
3. Monitor their condition over time. Consult a vet if needed.

Prevention

- Always feed a complete, balanced feed appropriate for their stage of life.
- Limit scraps and treats.
- Ensure access to clean water. Store food properly to avoid spoilage.

FATTY LIVER HEMORRHAGIC SYNDROME

When I pick up my chickens, I always make a quick note of their keel bone!

FATTY LIVER HEMORRHAGIC SYNDROME, or fatty liver, is a metabolic disorder characterized by excess fat in the liver and sometimes the abdomen. The enlarged liver can rupture, causing internal bleeding and sudden death. Early signs resemble those of other diseases, making diagnosis difficult. High-production hens, such as ISA Browns, Rhode Island Reds, and Leghorns, are more vulnerable because elevated estrogen promotes liver fat synthesis.

Causes

- Overweight birds with excessive fat on the abdomen and liver.
- Excessive calories or carbohydrates from treats or scratch.
- Being caged in a small space and being mainly sedentary.

Symptoms

- Pale combs and wattles.
- Excessive abdominal fat and poor body condition.
- Dull or ruffled feathers.
- Lethargy or reduced activity.
- Increased water consumption leading to loose stools.
- Poor shell quality or reduced egg production.
- Sudden death without previously seen signs.

Diagnosis

There isn't an at-home test for fatty liver hemorrhagic syndrome. A vet can use an X-ray, but diagnosis needs bloodwork, ultrasound, or liver biopsy—difficult for backyard chicken keepers. Signs are subtle and often mistaken for other illnesses, so prevention is best. To check whether your chicken is overweight, feel the front of their chest for the keel bone. A healthy bird's keel is not sharp and has normal muscle and flesh. An overweight bird's keel will be hard to find due to a thick fat layer.

Prevention

- Feed a balanced, nutritionally complete diet for the age of your flock.
- Avoid overfeeding by limiting chickens' scratch grain, treats, or scraps. The majority of their diet should be chicken feed, and the extras should stay under 10 percent.
- Ensure the flock has access to plenty of space for exercise. If the flock doesn't free-range or has no access to the outdoors, consider a large run with climbing structures and roosts.

GOUT

Sunny is a very large chicken. Oftentimes, she walks more slowly and has issues with swelling in her legs.

VISCERAL GOUT is a metabolic disease marked by high uric acid levels, where the kidneys can't excrete waste properly, causing crystals to deposit in organs or joints and leading to damage or death. It's common in older, nutritionally stressed flocks, and mainly as an acute form affecting young birds or entire flocks.

Articular gout is a chronic condition involving urate crystals around joints, affecting adult birds and often caused by diet, genetics, or mycotoxins. It impacts the feet and hocks, causing swelling, pain, redness, and mobility issues. Birds become reluctant to walk or roost.

Causes

- Incorrect mineral balance in the chicken's diet, such as calcium and phosphorus, or excess protein.
- Aging layers.
- High salt content in scraps or feed.
- Heat stress.
- Insufficient clean water and dehydration.
- Stress on the kidneys from toxins.

Symptoms

VISCERAL GOUT

- Sudden or gradual death in older hens, sometimes with no warning signs.
- Reduced egg production.
- Weakness, low energy, and decreased appetite.
- Poor water intake, tattered feathers, dehydration, lethargy, and weight loss due to kidney distress.
- Uric acid crystals found in the kidneys or organs after the bird has died.

ARTICULAR GOUT

- Swollen, enlarged, or deformed joints of the toes, legs, feet, or wings.
- Bumps on the joints from urate crystals.
- Lameness, reluctance to walk or rest, and occasionally standing on one leg.
- Reduced activity or eating.
- Weight loss over time.
- Permanent loss of function from joint swelling in chronic cases.
- Kidney function may remain somewhat normal.

Diagnosis

Diagnosing visceral gout in a live chicken is difficult because symptoms are mild and overlap with other illnesses. Postmortem necropsy reveals urate deposits or kidney lesions. Articular gout is easier to identify, indicated by lameness and swollen or deformed joints, with white urate deposits visible on tendons or joints.

Treatment

1. Provide supportive care, such as access to clean water, easy access to the coop and food, and reduced joint stress with low roosts, soft bedding, and ramps where needed.
2. Avoid high-protein or high-calcium diets, as those may cause further uric acid production.
3. Unfortunately, once urate crystals accumulate, the damage is likely chronic and irreversible. A vet could prescribe medications that may help reduce uric acid levels.
4. If the chicken is suffering, not responding to treatments, or is in terminal kidney failure, it may be best to consider euthanasia.

Prevention

- Feed a proper, balanced feed and avoid prolonged overfeeding of high-protein or high-calcium diets, especially in birds that do not lay eggs, such as young chicks and roosters.
- Always give your flock plenty of fresh, clean water. Dehydration can be a significant factor in the formation of urate crystals.
- Avoid contaminated feed or bedding that contains toxins or mycotoxins that could impair kidney function.
- Monitor for early signs of gout so you can adjust the feed and, hopefully, prevent further complications.

DIGESTIVE ISSUES

CHICKENS EXPERIENCE a range of digestive issues, including crop impaction, sour crop, and pendulous crop.

CROP IMPACTION

Ricky is showing off her full crop! The lump in the front of her chest is where she stores food before digestion.

CROP BINDING, also known as an impacted crop, occurs when material, such as long grass or straw, becomes lodged in the crop. The lodged mass prevents the normal passage of food into the digestive system. When the crop can't digest, food accumulates, digestion stops, and the bird becomes dehydrated and starved.

Causes

- Eating fibrous materials such as long grass, weeds, hay, straw, or bedding.
- Ingesting foreign objects such as string, plastic, metal, or twine.
- Overeating scratch, whole grains, or feeds without access to grit or water.

Symptoms

- Enlarged, hard crop that does not reduce overnight when food access is removed.
- Reluctance to eat.
- Reduction or absence of droppings.
- Signs of dehydration, such as lethargy and weakness.
- Signs of starvation, such as weight loss, poor body condition, and lethargy.

Diagnosis

Chicken crops vary based on diet and water intake. A water-filled crop feels like a balloon, while food and water feel like oatmeal. Overeating can make the crop hard and firm, which may indicate impaction. To check, withhold food overnight—the crop should be smaller by morning. If it remains full and hard, impaction is likely.

Treatment

1. Start with a gentle crop massage to help break up the blockage. Work from top to bottom.
2. Separate the chicken and provide only grit and water for 24 hours—no feed or treats. A small amount of vegetable oil may help, but give it slowly, one drop at a time, and allow the bird to swallow to avoid aspiration.
3. If the crop remains impacted after a few days, contact your veterinarian. Surgical removal may be necessary.

Prevention

- Avoid access to long grass clippings or tall weeds.
- Do not use hay in your coop. Use straw instead.
- Dispose of all strings, plastic, twine, or other materials.
- Provide grit at all times.

SOUR CROP

SOUR CROP, OR CROP MYCOSIS, is a digestive issue in chickens in which the normal bacterial flora of the crop becomes disrupted. This disruption allows *Candida albicans* (yeast) or other fungal and bacterial organisms to take over. This overgrowth leads to fermentation and inflammation of the crop wall. A sour crop can also be a secondary problem when a chicken is dealing with impaction or crop stasis. An untreated sour crop can become fatal, and the underlying causes need to be corrected.

Causes

- Blockage or motility issues, causing food to remain in the crop and ferment.
- Moldy feed, spoiled food or scraps, unclean water, or contaminated feed.
- Antibiotics or other medications that disrupt beneficial bacteria, allowing yeast and fungal overgrowth.
- Diets heavy in carbs, sugars, or fibrous plant material, causing slow digestion.
- Weakened immune system, another illness, or stress.

Symptoms

- Crop distension (when the crop feels soft and full long after they have eaten).
- Foul, sour, or "yeasty" smelling breath.
- Drooling or regurgitating foul-smelling liquid.
- Gurgling noises.
- White patches in the mouth if the infection spreads from the crop.

Diagnosis

To determine whether the chicken has a sour crop, first perform a crop fullness test. Check the crop first thing in the morning, before the chicken has access to food. If it is full and squishy in the morning rather than empty, that indicates a crop-emptying issue. If the crop is distended and you notice a foul odor from the mouth, the crop is likely sour. Foul odor, access to spoiled food, recent antibiotic treatment, and recent impaction are all factors that increase the likelihood of sour crop.

Treatment

1. Isolate the chicken and remove access to feed. Allow the crop to empty. If there is an impaction, the crop will not empty, and that issue must be dealt with first.
2. If the crop is still full or doughy after 12 to 24 hours without food, gently massage from top to bottom. Do this for a few minutes several times a day. Do not turn the bird upside down to force vomiting, as this can allow fluid to enter the lungs and cause aspiration.
3. If the chicken is dehydrated, breathing is distressed, or regurgitation persists, you should reach out to your vet.
4. After the crop has emptied, you can introduce water with probiotics and scrambled eggs. Do this for a few days, watching for signs of the sour crop returning. You can introduce grains or softened chicken feed if the sour crop has subsided.
5. If the chicken is still dealing with sour crop, your veterinarian can check for a deeper impaction or prescribe antifungal medications to clear the yeast infection.

Prevention

- Avoid moldy or spoiled feed and kitchen scraps.
- Keep waterers clean and ensure the flock has access to grit.
- Follow antibiotics with probiotics to restore good bacteria.
- Maintain a clean, dry coop and run.

ASCITES/WATER BELLY

Water belly

ASCITES, BETTER KNOWN AS water belly, is the buildup of fluid in a chicken's abdominal cavity, causing a swollen belly. It can also build up around the heart. Water belly isn't a disease but rather a symptom of another underlying problem. It can involve the heart, lungs, liver, metabolism, and reproductive system, and sometimes is a combination of more than one problem.

Causes

- Pulmonary hypertension causes the right ventricle to become overloaded, leading to fluid leaking into the body cavity.
- Liver disease such as fatty liver hemorrhagic syndrome, fibrosis, or liver tumors.
- Reproductive issues such as cystic right oviduct, tumors, or inflammation.
- Kidney failure.
- Cancer.
- Chickens bred for fast growth, such as meat birds, experience pressure on the heart and lungs because their bodies outpace their internal organs.
- Chickens fed a high-protein, high-sodium, or high-energy diet may be at greater risk of metabolic and circulatory strain.

Symptoms

- Swollen, soft abdomen that is filled with fluid.
- Belly area that can feel squishy and look distended.
- Labored breathing due to fluid compressing the organs.
- More time sitting and reluctance to move.
- Purple or bluish coloration on the combs and wattles.
- Struggle to stand upright due to excess fluid.
- Death.

Diagnosis

The physical signs listed above (swollen abdomen, respiratory problems, lack of movement) can be used to diagnose water belly. The exact cause is harder to determine, but a necropsy showing acetic fluid, dilation of the right heart ventricle, or congestion of the liver, lungs, and heart would provide firmer confirmation. Drained fluid color may help narrow down the cause.

Treatment

There is no cure for ascites. Once the fluid buildup is present and heart or lung failure is taking place, there isn't much to do but provide supportive care. Other diseases, such as cystic right oviduct, can be managed with fluid drainage and supportive care, and the hen can usually continue to live for some time.

1. Draining the fluid may provide temporary relief. It isn't a cure, but it can make the bird more comfortable for a short time.
2. Support the chicken by reducing stress, limiting activity, and ensuring easy access to food and water.
3. The quality of life should be evaluated to determine whether humane euthanasia is needed.

Prevention

- Choose birds who aren't predisposed to heart conditions. Avoid meat birds or broiler breeds and opt for slower-growing heritage breeds.
- Keep a clean coop with good ventilation and low-dust bedding. Keep the coop clean of droppings to avoid exposure to ammonia.
- Keep young chicks warm, as cold stress can increase the risk of ascites.
- Use a nutritionally balanced feed and limit salt intake and the number of treats.

ASK THE POULTRY DOC

DRAINING WATER BELLY

The fluid will almost certainly reaccumulate over time, but many hens can live a high-quality life for months or years if they are drained periodically when symptoms reappear. Whether dealing with ascites or a cyst, the relief procedure is similar, but the target for a cystic right oviduct is specific.

Ideally, use a ¼-inch (6 mm) 18-gauge or 20-gauge needle. If you only have longer needles, trim the plastic sheath so only ¼ to ⅜ inch (6 to 10 mm) of the needle is exposed. This prevents you from puncturing internal organs. Thoroughly clean the skin of the lower abdomen with antiseptic.

Target the area roughly 1 inch (2.5 cm) below and 1 inch (2.5 cm) to the right of the vent (avoiding the midline). If the chicken is standing, insert the needle perpendicular to the skin. If the chicken is sitting, insert the needle where the belly touches the surface, keeping it parallel to the ground. For ascites, insert at a shallow downward angle until you see a "flash" of fluid; it should drip freely. For a cyst, once you enter the fluid-filled space, it should drain naturally.

PENDULOUS CROP

A follower sent this image of her sweet hen with a pendulous crop.

PENDULOUS CROP is an abnormal condition in which the crop is chronically stretched beyond its typical muscle tone. It hangs low on the chest and swings like a pendulum. It is often filled with liquid or feed that doesn't digest properly, causing a reduction in normal digestion and secondary problems like malnutrition or sour crop. It can also be uncomfortable for the chicken.

Causes

- Prolonged overeating.
- Excessive water intake in a short period.
- Consumption of fibrous material or coarse feed.
- Full or partial impacted crop.
- Blockage in the gizzard.
- Offering only pelleted feed.

Symptoms

- An enlarged, flaccid crop that hangs low or swings when the chicken walks.
- Delayed crop emptying.
- Reduced food intake.
- Weight loss and poor overall condition.
- Possible sour crop.

Diagnosis

Diagnosis is usually based on observation, history, and a physical exam. A veterinarian can help you rule out other causes, such as sour crop or crop impaction.

Treatment

1. Provide the chicken with small meals with easily digested feed like chicken crumble. Provide grit, water, and crop massages to help the food move through.
2. Support the crop with a sling or "crop bra" that can be made or purchased. This will help support the chicken but may not cure the problem.
3. In severe cases, a veterinarian may need to treat infections or surgically remove the blockage. Mild cases may improve with supportive care, but severely stretched muscles cannot fully recover.

Prevention

- Avoid gorging situations or binge eating by keeping food available at all times using a gravity feeder.
- Feed chick crumbles and provide foraging opportunities rather than just pelleted feed.
- Provide the flock with coarse grit to aid digestion.

TOXICITIES AND POISONING

Moldy feed

"TOXICITIES" are health issues caused by exposure to harmful substances like moldy feed, toxic plants or foods, heavy metals, pesticides, and chemicals. These exposures can lead to acute or chronic illness. Moldy feed, often containing mycotoxins from fungi on grains, can suppress the immune system and cause serious health issues. Heavy metals are more commonly found in industrial areas, older buildings, or environments with scrap metal.

Causes

- Wet, moldy feed or grain.
- Poisonous plants or unsafe kitchen scraps.
- Heavy metals such as lead, cadmium, mercury, and arsenic found in the soil or water.
- Chemical contamination from pesticides, yard care, rodenticides, herbicides, and paint.

Symptoms

- Poor appetite or sudden drop in feed consumption.
- Diarrhea with unusual color, vomiting, or regurgitation.
- Weakness, lethargy, stumbling, and uncoordinated movements.
- Breathing difficulties (usually if they have inhaled toxins).
- Organ or systemic issues such as liver or kidney failure, poor growth, and increased susceptibility to disease.
- Sudden death without clear signs of illness.

Diagnosis

Diagnosing toxicity requires evaluating what the bird has been exposed to. Check for moldy feed, nearby chemicals or metals, and any questionable scraps or plants. Toxicities often affect multiple birds, causing poor performance or sudden death. Lab testing can confirm heavy metals or mycotoxins through blood, tissue, or feed analysis, and necropsy may reveal organ damage.

Treatment

1. Immediately remove the suspected toxin. Isolate the affected chicken, or chickens, so they are easier to treat.
2. Provide the birds with supportive care, including warmth and electrolytes.
3. Seek out veterinary care for possible chelation, fluid therapy, or other supportive measures.
4. Adjust the flock management to avoid future issues. Improve feed storage and evaluate the items fed to the flock.

Prevention

- Store feed in a dry, cool location. Discard moldy feed. Avoid dumping feed into large containers. Leave the feed in original packaging so residues don't mold and the feed is easily removed.
- Inspect the property for toxic plants and remove them if necessary.
- Avoid toxic human food scraps and processed food.
- Don't use rat poisoning, insecticides, pesticides, or lawn chemicals near the coop or where chickens could contact them.

CANCER AND TUMORS

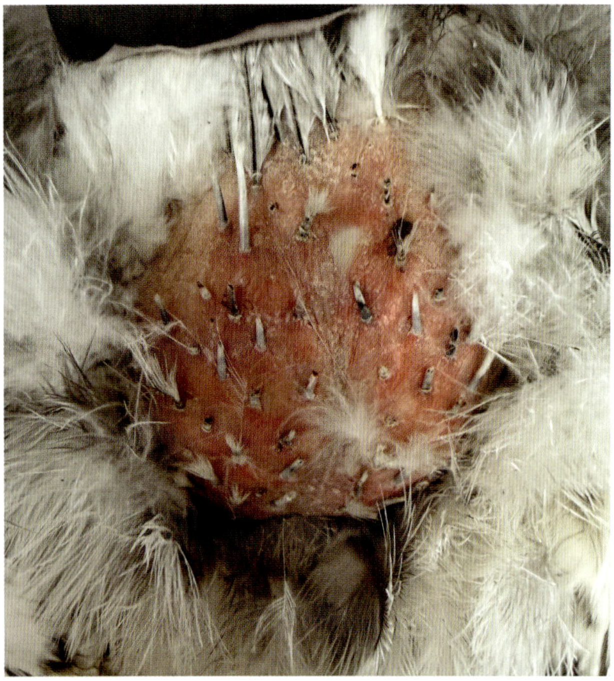

This chicken has a swollen abdomen, likely from a reproductive tumor or cystic right oviduct.

JUST LIKE OTHER ANIMALS, chickens can develop tumors that may be benign or malignant. Some tumors are induced by viruses common in poultry, such as Marek's disease and avian leukosis. Tumors can appear anywhere in and on the bird, including internal organs, lymphoid tissues, or skin and feather follicle tissues. These tumors, caused by abnormal cell growth, can behave differently depending on the cause, which can make diagnosis and outcomes vary.

Causes

- Viral diseases.
- Lymphomas, nerve and organ tumors, and sometimes skin or feather follicle tumors from Marek's disease.
- Lymphoid tumors, sarcomas, and cancers from leukosis or sarcoma viruses.
- Genetics or environmental factors.
- Older chickens or those with a weakened immune system can get tumors more easily.

Symptoms

- External or visible tumors that can be seen or felt under the skin, near feather follicles, or in the mouth.
- Lethargy, poor appetite, or weight loss.
- Abdominal swelling from fluid buildup or enlarged organs.
- Swollen joints, limbs, or face.
- Difficulty eating or swallowing.
- Poor egg production.
- Weakened immune system.
- Unexplained death.

Diagnosis

Tumors can mimic other chronic diseases, and oftentimes the symptoms are hidden until the tumor has progressed significantly. Tumors are often confirmed after death during a necropsy. To determine the cause and type of tumor, lab work would need to be completed.

Treatment

1. Tumors are challenging to treat, especially when they are malignant or viral in origin. Prevention is key.
2. Tumors often seriously worsen the quality of life. Humane euthanasia is sometimes the only practical option, especially if the chicken is in pain.

Prevention

- Prevent tumors with vaccination, when possible, good biosecurity, and avoiding mixing infected chickens with the flock.
- Avoid breeding birds that are known to develop tumors, likely due to genetic factors.
- Manage stress, minimize overcrowding, and avoid introducing sick birds.
- Maintain strict biosecurity.
- Monitor the flock to watch for early signs.

CONCLUSION

AS WE'VE SEEN IN THIS CHAPTER, internal ailments in chickens can quietly undermine flock health, productivity, and well-being. Many of these issues share subtle or overlapping signs, reinforcing the importance of careful observation, good husbandry, and prompt action when something seems "off." While no backyard keeper wants to think about sickness, understanding these internal processes empowers you to prevent problems before they escalate.

With that solid foundation in place, we now turn our attention to another critical aspect of chicken health: the egg-laying system. Reproductive and egg-laying ailments can take a chicken keeper by surprise, but recognizing what is happening is key to keeping your hens safe.

Unfortunately, a lot can go wrong during the egg-laying process. But when all goes well, the results are beautiful (and tasty!).

7

EGG-LAYING *Ailments*

Egg laying is the primary reason many people keep chickens. It is also one of the most complex biological processes a hen's body undertakes, which means it can come with many potential problems. As we move through this chapter, we'll explore the most common ailments and abnormalities that can arise within the reproductive system. From strange eggs to reproductive emergencies, these conditions can be clues to your hens' health. Clues seen on the outside are often indications of what's happening inside a hen's body. By understanding the signs, causes, and treatments of these conditions, you'll be able to implement preventive strategies and be better equipped to keep your hens healthy and productive.

SUPPORTING EGG LAYERS

MODERN CHICKENS TRACE their ancestry to a small, colorful wild bird called the red junglefowl. It was native to Southeast Asia and lived along forest edges and in jungle clearings. It foraged for seeds, insects, and small fruits. Red junglefowl were not initially valued for their eggs because, in the wild, they laid only a small clutch once or twice a year. The junglefowl, which laid ten to twenty eggs a year, is a big difference compared with chickens today, which lay two hundred to three hundred eggs.

Mabel wants to be a mother so bad she has collected more eggs than she can cover.

The red junglefowl became domesticated and used as a religious symbol, for cockfighting, as status animals, and as timekeepers due to the rooster's crowing at dawn. As time went on, people began selective breeding for hens that would lay more eggs and for hens that would lay regardless of fertilization. Humans rewired the chicken's reproductive system, so we must support them in their current stage of evolution.

Here are some basic ways you can support your flock's reproductive cycle:

1. **Prioritize Nutrition Over Production:** Egg formation requires protein, fat, vitamins, and calcium. A well-nourished hen can form strong shells and avoid depletion. High-quality layer feed supports egg production, but some hens benefit from additives. Provide free-choice calcium like crushed oyster shells separately so hens can take it as needed. Offer treats like black soldier fly grubs during recovery or heavy laying.
2. **Respect Natural Laying Breaks:** These include molting, seasonal slowdowns, illness, aging, and stress that reduce egg production for feather growth, rest, and repair. Supplementing with artificial light forces laying when recovery is needed. Avoid artificial light and accept fewer eggs during this state.
3. **Minimize Chronic Stress:** Stress hormones interfere with reproductive hormones, disrupting egg laying and increasing reproductive disorder risk. Maintain flock stability by providing space, keeping the coop and run clean, protecting from predators and weather, and monitoring health and parasites.
4. **Support Gut and Immune Health:** The reproductive tract overlaps with the digestive and immune systems. Chronic inflammation or gut imbalance can alter how a chicken lays, leading to internal laying, infections, and abnormalities. Keep their water clean at all times, use probiotics when there is gut upset, offer fermented feed, support their gut after harsh medications or antibiotics, monitor for internal parasites, and provide quality feed.

A hen's reproductive cycle was never meant to remain active for months or years at a time. Continuous egg laying strains the ovaries, oviduct, and calcium reserves, increasing the risk of reproductive disorders such as egg binding, prolapse, infections, and more. These conditions are not the hen's failure but rather a consequence of human-directed breeding for high productivity. As we walk through common reproductive ailments, understanding this history helps us chicken keepers recognize that many egg-laying ailments cannot be prevented with flock management alone.

ODD EGGS

ODD EGGS INCLUDE EGGS with thin shells, no shell, tails, bumps, multiple yolks, wrinkles, cracks and seals, and much more.

Thin-Shelled Eggs

These eggs form when the shell gland doesn't deposit enough calcium before laying, often due to calcium imbalance, vitamin D deficiency, stress, illness, heat, or in older hens or heavy producers. Avoid eating these eggs, as the thin shell may let pathogens in.

Shell-less Eggs

Shell-less eggs lack a hard shell, feeling rubbery because they consist only of the inner membrane holding the egg white and yolk. Common in new layers as their bodies adjust, they can also result from stress, fright, illness, or reproductive issues. Occasional shell-less eggs aren't concerning, but chronic ones may be caused by shell gland problems or infection and are unsafe to eat.

Eggs with Tails

While a surprising find, eggs with tails are not usually worrisome. Most of the time they are a glitch in the timing or the shell gland. They are often caused by stress, heat, fright, or just a young hen's body still figuring things out. These eggs shouldn't be eaten.

Bumpy Eggs

Bumpy eggs happen when excess calcium deposits unevenly on the shell due to shell gland irritation, infection, or an egg's prolonged time in the oviduct. They can also occur if a hen consumes too much calcium, which it needs to eliminate. These eggs are safe to eat.

Multiple Yolks

Multiple yolks occur when two or more yolks are released into the oviduct at the same time. This is most common in young hens whose reproductive cycles have not yet synchronized, though it can occur occasionally in mature hens as well. I have also seen it in my high-production breeds, which are selectively bred to lay a high number of eggs per year. These eggs are usually nothing to worry about and are safe to eat.

Wrinkled Eggs

Wrinkles occur when shell formation is disrupted by stress, illness, dehydration, or reproductive strain. The shell may partially collapse before hardening. These eggs often indicate a temporary disturbance but can also result from viral infections, a defective shell gland, nutritional imbalance, or environmental stressors. The occasional wrinkled egg is usually minor; however, if your hen consistently lays wrinkled eggs, you should investigate further. These eggs are safe to eat if the shell is intact and they smell normal.

Cracked and Sealed Eggs

These eggs, known as body-checked eggs, crack internally while still in the oviduct, then receive additional calcium that seals the damaged area. They usually form due to sudden movement, stress, or uterine contractions during shell formation. They are safe to eat as long as the shell is intact.

Egg Within an Egg

This rare and fascinating phenomenon occurs when a fully formed egg reverses in the oviduct and reenters the shell gland, becoming encased in a second layer of albumen and shell before being laid. Known as counter-peristalsis, it's usually a temporary glitch that poses no health issues unless the egg gets stuck. These eggs are generally safe to eat.

Fairy Eggs

Fairy eggs go by many other fun names, including wind eggs, witch eggs, rooster eggs, and fart eggs! They are tiny eggs that contain no yolk or only a tiny piece of one. They occur when a piece of tissue stimulates the oviduct to form an egg without ovulation. This happens more often in young pullets just starting to lay or in older hens near the end of production. They are nothing to be concerned about and are safe to eat, although not very filling!

Flat Eggs

Flat-sided eggs occur when the shell is compressed before it fully hardens. This can happen if the egg is crowded in the shell gland, for example, by another egg pressing against it. Muscular contractions can also distort the shape. Flat eggs are often linked to stress or timing issues, and they are safe to eat.

Misshapen Eggs

Misshapen eggs result from irregular muscle contractions in the oviduct or incomplete shell formation. They can be lopsided, elongated, or asymmetrical. They are more common in new or old layers and are safe to eat.

Bloody Shells

Blood smeared on the outside of the shell is due to small tears in the reproductive tract or vent. This is most common in young pullets laying their first few eggs. It can also occur when a hen lays an abnormally large egg or has irritated or inflamed tissue. It is usually temporary, and the eggs are safe to eat. It is always a good idea to observe the hens to make sure they don't have any external injuries from which the blood could have come.

Meat Spots

Meat spots are common in backyard chicken eggs. Commercial eggs are candled before sale, so imperfections are less likely to reach consumers. Commercial brown eggs are more likely to have meat spots because their darker shells make candling harder. The spots are small tissue pieces that break loose during egg formation. They are harmless but can be removed if preferred.

Blood Spots

Blood spots form when a small blood vessel on the yolk ruptures during ovulation. They are usually the result of minor reproductive trauma or a vitamin imbalance. They do not indicate fertilization or disease. They are safe to eat, though I cook them and feed them back to the chickens or give them to my dogs as a treat.

Watery Whites

Watery egg whites occur when the albumen lacks structure. This is common in aging hens, hens under stress, hens with poor nutrition, or improper storage of the eggs. Eggs left on the counter or in the coop too long can become thin and watery. If it isn't a storage issue, the eggs are usually safe to eat.

Sulfur/Rotten-Smelling Eggs

Eggs with a rotten, sulfurous, or sour odor are typically the result of bacterial contamination within the egg. This can occur due to a cracked shell, poor shell quality, or a reproductive infection. These eggs should be discarded.

Fishy-Smelling Eggs

A fishy odor is typically linked to diet. Chickens fed diets high in fish meal, fish oil, or omega-3s can produce fishy-smelling eggs. Treats such as cabbage or kale can also give eggs a temporary fishy odor. Very rarely, a genetic condition affecting choline metabolism can cause this. Fishy-smelling eggs are safe to eat but may be unpleasant.

Metallic- or Chemical-Smelling Eggs

A sharp chemical odor can develop when eggs absorb environmental odors. Check the coop for strong odors, such as ammonia buildup or cleaning chemicals. If you store your eggs near cleaning solutions, garlic, onions, candles, or air fresheners, they may also develop an off smell.

Musty- or Moldy-Smelling Eggs

Musty eggs are typically caused by moisture-related contamination. Eggs stored in a damp environment or put away while still wet can develop mold on or inside the egg. Poor ventilation in the coop can also contribute. These eggs should be discarded.

REPRODUCTIVE AILMENTS

EGG PRODUCTION IMPOSES UNIQUE and continuous demands on a hen's body. Modern chickens are biologically driven to lay frequently and for extended periods, unlike their ancestors, which laid only periodically. Many hens lay successfully for years, but not all do. Constant egg laying increases the risk of disorders and conditions affecting the reproductive organs.

Reproductive ailments are among the most common health issues in backyard flocks. These issues can range from mild and temporary to severe and life-threatening. Some develop suddenly, while others progress slowly and are easily overlooked until symptoms become severe. Understanding these disorders helps us recognize early warning signs, allowing us to prioritize both health and quality of life.

WHERE ARE ALL THE EGGS?

Finding an empty nest box can be frustrating, especially when your hens seem healthy and you've invested in feed. A sudden drop in egg production can have many causes, many of which are normal for healthy hens. Egg drop syndrome is a sudden decline or halt in egg production, usually due to stress, nutrition, seasonal changes, or reproductive fatigue rather than disease. Here are some possible reasons:

- Molting: Hens often stop laying entirely while they grow new feathers.
- Shorter daylight hours: Light reaching the bird's brain triggers hormones (from glands such as the pituitary and hypothalamus) that stimulate the ovary to produce eggs. Reduced light lowers hormone levels and naturally suppresses egg production, likely to conserve energy given limited foraging and the need to stay warm. This doesn't usually happen in a hen's first fall/winter, but it can.
- Stress: Predators, weather changes, flock disruptions, and overcrowding can contribute to decreased or halted egg production.
- Age-related slowdown: Older hens may lay sporadically or stop altogether.
- Illness or reproductive disorders: Internal laying, tumors, and salpingitis can result in decreased or halted egg production.
- Nutritional imbalance: Inadequate protein, calories, or calcium can affect egg production.
- Egg eating: Thin shells, overcrowded nest boxes, and a lack of soft bedding can cause eggs to break when laid. Chickens will eat these eggs even if they don't usually eat eggs from the nest!
- Hidden laying: Hens like to find hiding places for their eggs, especially if they free-range!
- Predators: Snakes, rodents, and other animals may sneak a snack from the nest box!

EGG BINDING

The penguin stance is a typical response to various reproductive conditions. It is most associated with the hen being egg-bound, but it can be a symptom of many other issues.

EGG BINDING OCCURS when a hen cannot pass an egg through the oviduct or vent. It is a medical emergency because it can quickly become life-threatening if the egg does not pass. Although egg binding is more common in younger hens, it can occur at any age. Without intervention, it can lead to internal rupture, infection, organ failure, and death.

Causes

- Calcium and vitamin D deficiency.
- Oversized, misshapen, or soft-shelled eggs.
- Dehydration.
- Obesity or poor body condition.
- Inflammation or infection in the reproductive tract.
- Stress.
- Forced egg production without recovery periods.
- Genetic predisposition or deformities.

Symptoms

- Constant straining or attempts to lay an egg.
- Penguinlike stance or tail-down posture.
- Swollen or firm abdomen.
- Reduced appetite, lethargy, or reluctance to move.
- Labored breathing and closed eyes in severe cases.
- Pale comb.

Diagnosis

Clinical signs and a physical examination are needed to diagnose an egg-bound chicken. If a chicken is showing any of the common egg-bound symptoms, you can gently palpate the abdomen and the back of the chicken to feel for a retained egg. Not all eggs are easily detected, especially when they are soft-shelled or shell-less. A vet could perform an X-ray or ultrasound to determine whether there is a stuck egg, but you can also feel with a gloved and lubricated finger. If the egg is lower in the system, you should be able to feel it.

Treatment

1. Your hen will need immediate care. Once a hen is egg-bound, the clock starts ticking. She has a day or two before it becomes life-threatening. Remove the hen from the flock to a warm location.
2. Calcium can help if the hen is egg-bound due to uterine muscle failure. Calcium is essential for smooth muscle contractions.
3. Lubricate the vent with a personal lubricant. Place the hen in a quiet, warm space where she can relax and focus on laying the egg. If the egg is still stuck, contact your vet, as it may need manual extraction. This is dangerous because a broken egg will very likely lead to infection or internal damage.

Prevention

- Nutritionally balanced feed with free-choice calcium on the side.
- Access to daylight for vitamin D.
- Constant access to water.
- Avoid obesity and chronic stress.
- Allow for natural breaks in the laying cycle.

ASK THE POULTRY DOC

CALCIUM SUPPLEMENTS

Give the hen a liquid calcium supplement, such as liquid calcium gluconate. She should receive 0.5 to 1 ml orally. Do not exceed more than two doses without the guidance of a veterinarian.

INTERNAL LAYING

INTERNAL LAYING OCCURS when a hen releases yolks from the ovary, but the yolks fail to pass properly through the oviduct. Instead of forming a shell and being laid, the yolks are released into the abdominal cavity, causing inflammation and serious side effects.

Causes

- Reproductive tract damage.
- High egg production.
- Hormone imbalance.
- Poor muscle function.
- Reproductive decline due to age.
- Genetic predisposition.
- Stress or illness.
- Calcium deficiency.

Symptoms

- Gradual abdominal swelling that worsens over time.
- Reduced or stopped egg production.
- Decreased appetite, lethargy, and decreased activity.
- Penguinlike stance, tail-down posture.
- Pale comb.

Diagnosis

The abdomen will feel fluid-filled or like a soft mass. Although the symptoms resemble those of being egg-bound, egg-bound chickens do not live long without passing an egg. If your hen is experiencing these symptoms and is still alive, there is a good chance she is internally laying.

Treatment

1. Although there is no cure for internal laying, supportive care may help the chicken feel better. A vet can prescribe an anti-inflammatory medication, hormone therapy to stop ovulation, fluid removal, and pain management.
2. In severe cases, humane euthanasia may be the best option to save the chicken from suffering.

Prevention

- Avoid artificial light that forces egg laying.
- Ensure balanced nutrition and supplemental calcium.
- Minimize stress.
- Avoid high-production breeds.

EGG YOLK PERITONITIS

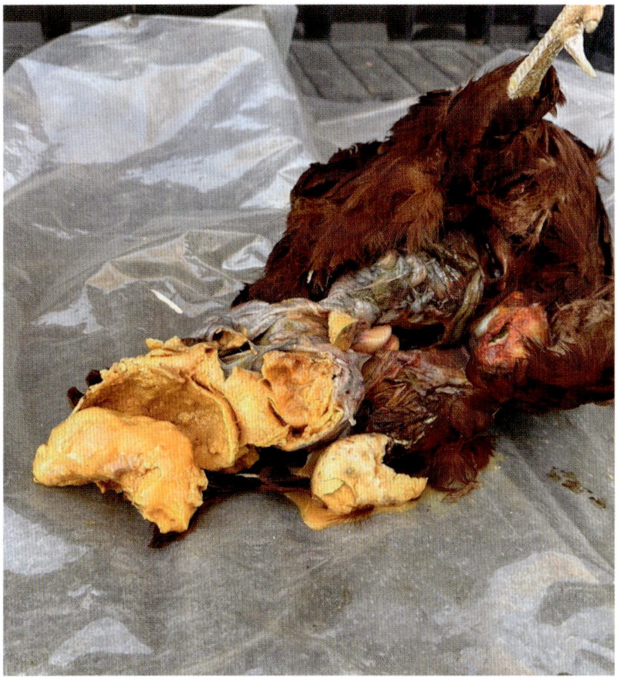

This hen suffered from egg yolk peritonitis, and when a necropsy was performed at home, my follower found old egg yolk material inside her abdominal cavity.

EGG YOLK PERITONITIS occurs when yolk material enters the abdominal cavity, triggering inflammation and infection. Yolks in the oviduct are usually sterile, but once they are in a new location, they provide a medium for bacterial growth. This condition is life-threatening because it can lead to infection and sepsis.

Causes

- Internal laying.
- Reverse peristalsis, when muscle contractions push an egg backward from the oviduct into the abdomen.
- Egg binding, reproductive tumors, and salpingitis.
- High stress, weakened organs of an aging hen, and excessive artificial light at a young age.
- Oviduct damage or infection.

Symptoms

- Abdominal swelling or distention that worsens over time.
- Decreased appetite, lethargy, and weakness.
- Pale comb, labored breathing.
- Tail-down, penguinlike stance.
- Decline in egg production.
- Fever or other signs of sepsis.

Diagnosis

The condition is usually advanced by the time it is recognized. A definite diagnosis may require a veterinarian imaging and fluid sampling. A yellowish or egglike discharge from the vent may occur.

Treatment

1. Egg yolk peritonitis can be treated, but it is rarely cured unless the hen is forced to stop laying eggs. This is because there is likely an underlying issue causing the peritonitis that will keep happening. Because hens will hide symptoms until the infection is advanced, the prognosis is frequently poor.
2. Mild cases without a secondary bacterial infection may resolve with isolation, warmth, high-quality nutrition, and anti-inflammatory medications.
3. More advanced cases need veterinarian intervention. Antibiotics and drainage may be used. The vet may use hormonal implants or surgery to permanently stop the hen from laying eggs.

Prevention

- Avoid forced egg laying with artificial lighting.
- Provide balanced nutrition and free-choice calcium.
- Minimize stress.
- Avoid breeding from hens with reproductive disorders.
- Use gentle handling, as rough handling can cause an egg to break internally.

SALPINGITIS

SALPINGITIS IS AN INFLAMMATION and infection of the oviduct. It is a serious, often progressive condition that not only disrupts egg laying but can also lead to chronic reproductive failure. It can occur independently or as a complication of another reproductive problem. Once the oviduct becomes inflamed, its ability to move yolks, secrete albumen, and form shells is compromised.

Causes

- Bacterial infection introduced through the vent, especially in dirty conditions.
- Egg binding or retained egg material.
- Internal laying.
- Internal egg rupture.
- Chronic frequent egg laying without natural breaks.
- Stress or immune suppression.
- Poor shell quality that causes internal breaks.
- Weak reproductive tissue due to age or illness.

Symptoms

- Drop in egg production.
- Soft-shelled, shell-less, or oddly shaped eggs.
- Eggs with foul odor or abnormal contents.
- Decreased appetite, lethargy, and reduced activity.
- Abdominal swelling.
- Tail-down, penguinlike stance.
- Droppings soiled around the vent.
- Fever or other signs of systemic illness.

Diagnosis

Because salpingitis is often misdiagnosed as internal laying or egg yolk peritonitis, it requires a veterinarian's examination for confirmation. Diagnostic approaches include physical examination, imaging, abdominal fluid sampling, and, if indicated, a culture of the discharge.

Treatment

1. Isolate the chicken from the flock and provide supportive care. This may include hydration, nutrition, and warmth. Use electrolytes if the chicken is weak.
2. Acute cases caught early may respond to treatment that controls infection, reduces inflammation, and manages ongoing ovulation. A vet may prescribe antibiotics, anti-inflammatory medication, and hormonal therapy to suppress egg production.
3. More advanced cases may progress despite interventions. Long-term management may focus on comfort and quality of life.

Prevention

- Maintain a clean, dry environment.
- When a hen has droppings accumulated on her, they should be cleaned promptly.
- Provide clean water and quality nutrition.
- Minimize stress and overcrowding.
- Refrain from using artificial light.
- Address reproductive issues as they occur to avoid infection.

REPRODUCTIVE TUMORS

REPRODUCTIVE TUMORS are abnormal growths in the ovaries or oviducts of laying hens, common in high-production and older hens. Associated with long-term, repetitive laying, these tumors can be benign or malignant, and even benign ones can impair reproductive health. They develop slowly and silently, often going unnoticed until large. Types include ovarian, oviduct, mixed reproductive tract, and cystic masses.

Causes

- Age-related cellular changes.
- Genetic predisposition.
- Hormonal stimulation.
- Chronic egg production without proper rest periods.
- Inflammation or other diseases.
- Breeding for high egg output.

Symptoms

- Progressive abdominal swelling.
- Reduced or absent egg production.
- Weight gain.
- Lethargy, decreased activity, tail-down posture.
- Pale comb, labored breathing.
- Fluid accumulation in the abdomen.

Diagnosis

To make a definitive diagnosis, a veterinarian would perform an ultrasound or X-ray imaging, analyze any ascitic fluid, and conduct abdominal palpation. Unfortunately, by the time problems are visible, the tumors are usually far advanced.

Treatment

1. Treatment for reproductive tumors is difficult. There is no feasible cure for tumors because surgery comes with too many risks to be successful.
2. Focus on quality of life and pain management. A vet may prescribe pain and anti-inflammatory medications. Hormonal therapy to halt egg production and periodic drainage of abdominal fluid may also be needed. It is important to discuss quality of life with your vet to determine whether humane euthanasia is necessary.

Prevention

- Avoid artificial lighting that forces continuous egg laying.
- Support natural egg-laying breaks.
- Provide clean water and balanced nutrition.
- Minimize overcrowding and stress.
- Avoid high-production breeds and avoid breeding hens with reproductive disorders.

VENT GLEET (CLOACITIS)

When I see one of my chickens with matted feathers like you see here, I jump into action! Leaving the feathers dirty can cause so many problems.

VENT GLEET IS AN INFLAMMATION of the cloaca, the exit for droppings and eggs. It's often linked to yeast overgrowth, with possible bacterial causes. Not a reproductive disease, but more common in hens due to moisture, hormones, and reproductive strain, it occurs when cloacal microorganisms are disrupted. Left untreated, it can damage the vent or lead to secondary infections.

Causes

- Yeast overgrowth following antibiotic use.
- Imbalance in pH, which allows harmful organisms to thrive.
- Bacterial infections like *E. coli* or Salmonella.
- Spoiled feed.
- Chronically moist or dirty vent feathers.
- Poor ventilation or damp living conditions.
- Poor nutrition, feed high in sugar or carbohydrates.
- Diarrhea that causes a buildup of bacteria.
- Stress, illness, or a weak immune system.
- Mating by a rooster who has mated with another hen dealing with vent gleet.
- Trauma from laying unusually large eggs.

Symptoms

- Telltale yellowish-white, sticky discharge coming from the vent.
- Strong odor.
- Soiled vent and tail feathers.
- Inflammation of the vent that can appear red, swollen, and irritated.
- Soft and bloated abdomen, later becoming hard to the touch.
- Decreased egg production.
- Lethargy or discomfort, especially when the condition is advanced.

Diagnosis

A visual inspection of the vent and surrounding area is usually sufficient for diagnosis. Hens can have droppings caked onto their feathers for many reasons. To diagnose vent gleet, look for yellowish discharge, soiled vent feathers, and a foul odor that is more yeasty than rotten.

Treatment

1. Start by cleaning the vent feathers. The easiest way I have found is to soak the chicken in some warm, soapy water before working on the feathers with your hand. Use a glove if you don't feel comfortable using your bare hand. This can take some time, but I prefer not to trim the feathers unless I have to. Trimmed feathers can leave skin bare and more likely to get picked at by others in the flock.
2. Treat the fungal infection with an antifungal medication. Options include topical and oral treatments. You can consult your veterinarian or try a human yeast infection medication. Add probiotics to their water and refresh it daily until gut balance is restored.
3. Make adjustments to the environment and nutrition as needed. Some changes might include improving coop cleanliness and ventilation, reducing excess sugars and carbohydrates in their scraps, and ensuring they aren't overcrowded.
4. Oftentimes, vent gleet requires ongoing management and isn't a onetime fix. Recurrent cases are common if the underlying cause persists.

Prevention

- Tend to dirty vents when spotted instead of letting them stay dirty.
- Maintain a clean coop with good ventilation.
- Avoid excessive sugary treats.
- Support gut health with balanced nutrition and, when needed, probiotics.
- Avoid antibiotic use without a vet's guidance.

ASK THE POULTRY DOC

VENT GLEET

To effectively treat vent gleet (cloacitis) in backyard flocks, a data-driven protocol prioritizes correcting the internal urogenital environment rather than merely scrubbing surface symptoms. The primary objective is to restore the cloaca's "acid shield" by lowering the pH to an inhibitory range of 6.0 to 7.0, which effectively stalls the growth of opportunistic pathogens like *Candida albicans*.

The use of acidified copper sulfate (ACS), dosed at approximately ¼ teaspoon (1 ml) per 1 gallon (3.8 L) of drinking water for 3 to 5 days, acts as both a fungistat and a bacteriostat. Simultaneously, keepers must address the "chemical load" of the diet; if the bird's chemistry is out of balance, reducing calcium intake to less than 3 percent can help resolve suspected metabolic alkalosis and restore normal urogenital chemistry.

For local wound management, daily cleaning is essential to remove the caustic urates that cause painful urine scald. Once the area is clean and dry, a safe barrier ointment—such as silver sulfadiazine, diaper cream, or plain petroleum jelly—should be applied to protect the tissue. Crucially, keepers must strictly avoid steroid creams, as they suppress local immunity and can exacerbate the underlying infection, whereas topical antifungals can serve as a helpful support to the systemic therapy provided by the ACS.

PROLAPSED VENT

A follower sent me this image of her chicken's vent prolapse. The prolapse seems severe, but she was able to make a full recovery.

THE VENT IS THE CHICKEN'S OPENING for droppings and eggs. A vent prolapse occurs when internal tissue protrudes through the vent, often affecting laying hens and exposing the tissue to injury, infection, and pecking. It's a medical emergency, with prognosis depending on severity, cause, and response time. Mild cases may heal, but repeated prolapses are harder to manage and harm the chicken's health. Tissue damage and infection impact outcomes; quality of life should guide decisions.

Causes

- Excessive pressure or weakened tissue in the reproductive tract.
- Laying large eggs frequently.
- Forced laying with artificial light before a hen is mature.
- Egg binding, internal laying, or salpingitis.
- Obesity and poor muscle tone.
- Calcium deficiency.
- Inflammation or infection.
- Constipation.
- Genetic predisposition or early-onset laying in young pullets.
- Previous prolapse makes it more likely to happen again.

Symptoms

- Pink, red, or swollen tissue protruding from the vent.
- Straining or repeated attempts to lay an egg.
- Bleeding, tissue discoloration.
- Soiled vent area.
- Lethargy and distress.
- Pecking and attention from flock members.

Diagnosis

Diagnosis is based on visual examination. Protruding tissue indicates a vent prolapse, but you must also assess severity and possible cause. When examining the chicken, note whether the tissue is moist, swollen, or damaged. Determine whether an egg is present and needs removal to help the tissue retract. For severe cases or those you are uncomfortable handling, consult your vet.

Treatment

1. Isolate the chicken to prevent pecking from the rest of the flock. Soak the chicken in a warm bath for 20 minutes. Clean the vent area of blood, debris, and droppings.
2. Assess the damage. If the tissue is damaged from pecking, proceed gently. The damaged tissue will need to be monitored for infection and kept clean. I prefer to use a chicken wound care spray a few times a day.
3. Lubricate the vent with coconut oil or a personal lubricant to keep it moist while you work on the prolapse. Using a gloved finger, feel inside for a stuck egg or droppings. Help the chicken pass the egg by gently guiding it out. Remember, a broken egg is likely to lead to infection, so work slowly. If you find dried or stuck droppings, use the lubricant to help clear them away.
4. Once the pathway is clear, it is time to guide the prolapsed tissue back into its proper position. Gently work the tissue back into the chicken. I recommend holding it in place with your hand or a clean, warm towel for 15 minutes. The risk of it prolapsing again is high, so be patient.
5. Once the prolapse is back in place, put your chicken in a warm, dry location, away from the flock. They will need to stay here for a few days. I advise not using bedding as the bedding can stick to the prolapse. Use puppy pads or clean towels and change them out often. The goal now is to temporarily stop egg production while they heal. Limit their access to light by keeping them in the dark for most of the day and night.
6. Check on them a few times a day, and repeat the lubrication and repositioning of the prolapse. Hold it in with your hand for 15 minutes and place them back in their cage. This process can take several days, and longer if their tissue was damaged.
7. Contact a veterinarian if the tissue is discolored, dry, infected, or severely damaged.

Prevention

- Avoid artificial lighting that forces egg laying.
- Support muscles and shell quality with a quality calcium source as a free-choice option.
- Avoid extra treats that could cause obesity.
- Monitor the flock for signs of prolapse to catch it early.

CYSTIC OVARIES

CYSTIC OVARIES OCCUR when ovarian follicles fail to ovulate normally and instead develop into fluid-filled cysts. These cysts are not cancerous, but they can disrupt hormones and interfere with egg production. Cystic ovaries usually develop due to ovulatory failure, in which follicles mature but fail to release their yolks. They can significantly impact a hen's health and quality of life.

Causes

- Hormonal imbalance.
- Chronic high egg production.
- Genetic predisposition.
- Reproductive decline due to age.
- Stress.
- Previous reproductive disease.

Symptoms

- Reduced, irregular, or nonexistent egg laying.
- Abdominal enlargement.
- Lethargy, reduced appetite, lack of activity.
- Pale, sometimes enlarged comb.
- Persistent broody behavior.

Diagnosis

Symptoms are gradual and may fluctuate over time, making cystic ovaries hard to diagnose. Usually, a vet must confirm the diagnosis with abdominal palpation, ultrasound imaging, and ruling out tumors, ascites, or internal laying. Without imaging, diagnosis is nearly impossible. Oftentimes, a diagnosis is made after death during a necropsy.

Treatment

1. There is no cure for cystic ovaries, so management focuses on supportive care.
2. A vet may prescribe anti-inflammatory medication and monitor the size of the abdomen over time.
3. Provide your hen with a comfortable living space, good nutrition, and a low-stress environment.
4. Monitor the quality of life and discuss humane euthanasia if needed. Some hens remain stable for months while others progress and decline quickly.

Prevention

- Avoid artificial lighting that forces egg laying, and allow natural breaks in the cycle.
- Provide balanced nutrition.
- Minimize overcrowding and stress.
- Avoid breeding hens with this issue and passing on genetics.

CONCLUSION

EGG LAYING IS ONE OF THE MOST demanding functions of a hen's body. Many of the ailments discussed in this chapter are not the result of poor care but of the strain placed on the reproductive system, shaped by generations of selective breeding. Understanding these conditions helps chicken keepers recognize early warning signs, respond intentionally, and make informed decisions rooted in compassion rather than urgency.

As the focus shifts from adult hens to the earliest stages of life, the challenges change, but the goal remains the same. Proper care, observation, and support in a chicken's earliest days lay the foundation for lifelong health. Strong, healthy flocks are built early, and the next chapter explores how proper chick care sets the stage for everything that follows.

There's nothing better than home-raised eggs!

Chicks are cute but complex creatures. Giving them the right care is essential.

8

THE SMALLEST MEMBERS of the Flock

Baby chickens bring excitement to families with their peeps, fluffy feathers, and curiosity, making them favorites among many chicken keepers. However, raising chicks isn't for everyone. I was surprised by this, having grown up obsessed with baby animals and hoping orphaned ducklings or baby squirrels would show up so I could care for them. I always wanted chickens, but never thought of starting with adults. Caring for a baby animal (or human) is my happy place.

Caring for chicks requires attention to their needs, which differ from those of mature birds. Small mistakes can lead to major problems if not addressed early. Success begins when they arrive. Chicks are cute for about 2 weeks before becoming scraggly, dusty, noisy chaos.

Chicks need extra time for cleaning, dusting, training, feeding, and checking on them. They require specific equipment that older chickens don't. However, they're easier to bond with, especially for beginners. Many animal lovers find full-grown chickens intimidating. Starting with chicks helps you get used to them gradually and build a bond. While most chickens don't seek affection, early handling is important for caring for them when they're sick or injured.

I've discussed many ailments, injuries, and issues you might encounter in adult chickens. Chicks come with their own set of potential problems you'll need to know how to handle. Luckily for you, I've been hatching chicks for more than twenty years, and I believe I've seen nearly everything!

AVOIDING THE MOST COMMON CHICK MISTAKES

MANY NEW FLOCK OWNERS unknowingly make choices that put their chicks at risk. Understanding the most common pitfalls can spare you heartache later. One of the biggest mistakes is placing chicks in a brooder that is too cold or drafty. Chicks rely entirely on external heat for the first few weeks of life. If temperatures dip too low, even briefly, it can weaken their immune systems and cause chilling or death.

This type of waterer provides fresh, clean water to the chicks all day long.

Another common mistake is using a waterer that is too deep or unstable. Young chicks can drown if they fall into open water. There are two main types of waterers for chicks: an open gravity-fed waterer and a nipple waterer. If you use the gravity-fed type, add small marbles or pebbles to the tray to prevent drowning. The nipple waterer works when chicks peck at the spout, and water comes out. This is my favorite type because it doesn't get dirty. Make sure it is properly secured so chicks can't knock it down. If chicks get wet, they could catch a chill and die.

Overcrowding is another overlooked danger. A brooder that's too small can lead to piling, injuries, and even smothering at night, when chicks instinctively cluster together. It's better to start with more space than you think you'll need. If they're overcrowded, they can get sick from the buildup of droppings in the brooder. I suggest at least 1 square foot (0.1 square m) of space per chick to start. They will need more space the longer you keep them in the brooder.

Finally, it's important to avoid slick, smooth brooder flooring, such as newspaper or plastic. Chicks need good traction to develop strong legs and prevent conditions like splayed legs. You should opt for textured materials such as puppy pads or hemp bedding instead.

BROODER HEAT I like to think of brooder heat in terms of how a mother hen raises a chick. If you've never seen it, it may be hard to visualize. Mother hens provide all the heat their babies need. Chicks burrow under her feathers when they are cold, then come out into the warmth to eat, drink, and interact with their environment. Chicks need heat, but it doesn't need to be constant. Having space to get out of the heat helps them acclimate to outdoor temperatures much faster.

The chicks stay warm in their mother's feathers but come out when they want to explore.

A SUCCESSFUL BROODER

A BROODER IS THE CHICKS' entire world for their first several weeks, so a thoughtful setup is crucial. Start with a sturdy container, such as a large plastic tote, a livestock trough, a small animal cage, or a specialized brooder box. Make sure the walls are high enough to keep adventurous chicks inside but still allow easy access for you. I always suggest a breathable lid. Curious dogs and cats will make their way into the brooder if given the chance, and those chicks will be jumping out by week two if they aren't contained.

A comfortable brooder box will make raising chicks so much easier.

Heat is essential for a good brooder. Use a safe, adjustable heat source like a brooder plate or secure heat lamp. During the first week, keep the warmest area around 95°F (35°C), lowering it by 5°F (2.8°C) weekly as chicks grow. Provide unheated space for self-regulation. I prefer brooder plates over heat lamps due to fire risk and their similarity to a mother hen's warmth. Keep the plate low enough for chicks to squat under, but not so low it touches them. In the first 2 weeks, keep the brooder in the house because the plate warms only the bottom, unlike a heat lamp that heats a larger area.

These chicks are peeking out of their brooder plate, where they stay warm.

Feeders and waterers should be designed for chicks and secured so they don't tip. The biggest challenge is keeping bedding and droppings out of the food and water. Use only chick crumbles in the feeders and offer chick grit on the side. Check the food and water daily, as these are their sources of energy and growth. If your chicks are chirping quietly and exploring, you are doing a great job!

CHICK TIP An important step in managing a brooder is cleaning it out and reaching all the chicks. Make sure you can physically reach all parts of the brooder and that you can move it if needed.

These sweet brooder buddies are out on an adventure. It is okay to take chicks away from their heat source for short periods of time.

For the first few days, I like to use absorbent liners in my brooders. These can be unscented puppy pads. I will stack about four liners in the brooder and remove one in the morning and one at night. I do this to make sure the chicks have a few days to find their food. After I am sure they are eating, I will switch to bedding. There are many options for brooder bedding. I used to use pine shavings, but now I prefer hemp bedding. Hemp is very absorbent and removes the moisture from their droppings. I will remove the bedding once a week and top dress it as needed every few days. If you are using less absorbent bedding, you should remove it more often to ensure the chicks have a clean space.

Over the years I have learned that the setup makes all the difference. If you are finding things difficult to manage, you may need to make adjustments. Creating a comfortable, safe brooder from the start not only promotes healthy development but also gives your smallest flock members the best possible beginning to a thriving life in your backyard or on your farm.

Bedding is not only crucial for comfort, but it also absorbs the moisture in the brooder.

FEEDING CHICKS

MANY PROBLEMS ARISE when chicks aren't given adequate nutrition. There are many options and brands to consider, and it can be confusing, so knowing the basics will help you make an informed decision when choosing their feed. Feed companies categorize their feed by age and type of chicken. They also offer different physical forms of the feed. Here are the terms you should know when choosing chick feed.

Feed Consistency

- **Crumbles:** This is feed that is formed into small, irregular-shaped pieces, making it easy for young chickens to pick up and digest. You can feed crumble form to chickens of all ages.
- **Pellets:** This is feed that is formed into bigger pieces. There is less dust than with crumbles, but it is meant for older chickens. Stay away from this until they are laying age.
- **Mash:** Mash is the term used for whole grains that have been crushed into pieces. You can actually see the ingredients in the feed instead of it being formed into uniform pieces like you see with crumbles and pellets. The nutrients are still there, but chickens will often pick out their favorite pieces and leave a lot of powder, which is too hard for them to pick up.

Feed by Age

- **Starter Feed:** This feed is for chicks ages 0 to 8 weeks. It contains 18 to 20 percent protein to support rapid growth, higher levels of essential amino acids (such as lysine and methionine) to support muscle and tissue development, and lower calcium to avoid kidney damage.
- **Grower Feed:** This feed is for ages 8 to 20 weeks. The protein level is usually between 15 and 18 percent to support growth, but it is lower than that of fast-growing baby chicks. The vitamins, minerals, and amino acids are adjusted for chicks who are larger but not yet laying.

Other Terms

- **Medicated:** Some feed companies offer feed with a medication called amprolium added in to help prevent coccidiosis.
- **Chick Grit:** This is ground-up rock, usually granite, that chicks need to help grind and digest their food. They do not usually need it if they are only eating crumbles, but if they are eating mash or any treats, you'll want to provide it in a separate feeder.
- **Scratch Grain:** This is a mix of grains and treats that is not meant for chicks. It is nutritionally unbalanced and, even when they are older, should be used sparingly.

I suggest feeding your new chicks a starter crumble or mash feed for their first 8 weeks. You can choose medicated or unmedicated. If they are vaccinated for coccidiosis, choose unmedicated, as they are already protected and do not need both. After they turn 8 weeks old, you can switch to a grower feed until they are 20 weeks or until they start laying eggs, at which point you will use a layer feed.

The chicks will need a crumble or mash feed to start. The small pieces are perfect for them to pick up.

A Note About Homemade Feed

Feeding chicks homemade, unbalanced feed can cause growth and health issues that affect long-term productivity. Problems include skeletal deformities, stunted growth, low energy, digestive and immune issues, and future egg-laying problems. While many people share homemade feed recipes online, please be cautious when following them. Saving money is not worth your chicks' health. Feed companies employ nutritionists who determine which ingredients will best serve your chicks.

CHICK TIP Stay away from treats for a few weeks until they are eating well and growing. After that you may add in healthy treats now and then, but keep it to a tiny bit, as they need their balanced feed to grow properly.

COCCIDIOSIS

Blood in the droppings is a clear sign that coccidiosis has taken hold.

COCCIDIOSIS IS A PARASITIC disease of chickens and other poultry species caused by coccidia, protozoan parasites. These parasites infect the intestinal tract, causing inflammation and damage to the intestinal lining. It is one of the most common diseases in young chicks, especially those under 6 weeks of age, though it can affect older birds as well. Chicks can develop coccidiosis even without going outside or being exposed to other birds.

Causes

Coccidia are microscopic parasites that live in the intestines of chickens and chicks. As chicks age, they develop resistance to coccidia and can keep it under control. However, when they are young, their immune systems are still developing, making them more susceptible to an overload of coccidia, which can lead to coccidiosis. The infection is more likely when chicks ingest coccidia oocysts (the egglike form of the parasite) from contaminated bedding, feces in the brooder, and dirty feed and water. If the brooder is dirty with droppings, chicks can become infected faster. Here are the main causes:

- **Age:** Young chicks are more susceptible because their immune systems are still fragile and developing. Older chickens can also get coccidiosis, but it is less common.
- **Poor sanitation:** Chicks who are overcrowded or living in damp, dirty brooders are more likely to get coccidiosis.
- **Dirty water:** The most common source for coccidia to be found is in the waterer. Consider using a waterer that requires the chicks to tap the spout for water, as these stay clean and the chicks cannot kick dirty bedding into them.
- **Environmental stress:** Believe it or not, stress takes a toll on the immune system and can make chicks more susceptible to coccidiosis. Stressors include temperature fluctuations, diet changes, predator attacks, or any other stressful conditions.

Symptoms

- Diarrhea is usually the first symptom, appearing tan, green, or yellow, and can become bloody if intestinal walls break down. Blood in droppings indicates advanced infection needing urgent treatment. Note that chicks and adult chickens may shed pinkish intestinal lining, which is normal. Coccidiosis damage, however, shows as fresh red blood.
- Chickens infected with coccidiosis may appear lethargic, have difficulty standing, or sit with their wings drooping. They may eat less or stop eating altogether and become dehydrated.
- Damage to the intestinal lining impairs nutrient absorption, leading to malnutrition, weight loss, and poor growth in young birds.
- Infected chickens may stand around with their eyes closed and not want to move. They may also exhibit ruffled feathers and look puffed up.
- If untreated, coccidiosis can cause severe dehydration and organ failure, quickly leading to death.

Diagnosis

Coccidiosis can be detected by a veterinarian using a fecal test, but if you notice blood in the droppings, the test may take too long, and it may be too late. If the chick or chicken has already died, a necropsy can reveal intestinal damage, suggesting coccidiosis. If you do not have access to a veterinarian who treats chickens, fecal samples can be mailed to a lab. (See the Resources section in the back of the book.)

Treatment

1. To treat minor infections with just runny droppings, I've had success using oregano essential oil. I add a toothpick dip of oil into their waterer or mix it into fermented feed once a day. Continue for a few days until symptoms have subsided. I then follow with probiotics.
2. If the chicks have a more severe infection and are showing signs beyond runny droppings, such as blood in the stool, lethargy, and refusal to eat, I suggest moving to medication as soon as possible. Consult with your vet for instructions, but this is what has worked for me: I use medications such as amprolium or sulfonamides. You can find these in the cattle section of most farm stores. I prefer amprolium, but you must use the medication itself, not medicated feed. Medicated feed does not contain enough to treat an infection.

Prevention

- Use medicated feed for the first few weeks of life.
- Provide good sanitation.
- Vaccinate newly hatched chicks.
- Rotate foraging grounds to avoid dropping accumulation.

ASK THE POULTRY DOC

DOSAGE SUGGESTIONS

Coccidiosis is a common, stressful, and often misunderstood disease in backyard poultry. Controlling it through a "management over sterilization" approach, which prioritizes preventing death while allowing the bird's immune system to develop necessary resistance, is the goal.

- **High Dose** (0.024 percent/2 teaspoons [10 ml] per 1 gallon [3.8 L] of water) **for 3 Days:** Used during severe outbreaks with mortality or bloody droppings to rapidly crush parasite reproduction.
- **Medium Dose** (0.012 percent/ 1 teaspoon [5 ml] per 1 gallon [3.8 L] of water) **for 3 to 5 Days:** Recommended for moderate symptoms (diarrhea, ruffled feathers) without mortality, providing significant suppression of replication.
- **Low Dose** (0.006 percent/0.5 teaspoon [2.5 ml] per 1 gallon [3.8 L] of water) **for 3 Days:** Administered as a preventive "pulse" during high-risk periods (e.g., stress or wet weather) to reduce shedding while allowing the cycling of low levels of coccidia, which is the gold standard for maintaining flock immunity.

PASTY VENT/BUTT

This chick has pasty butt and is in need of cleaning right away.

PASTY VENT, more commonly called pasty butt, is very common in young chicks. It occurs when fecal matter accumulates on their vent (the opening through which they excrete waste). The droppings dry and stick to their back ends, making it impossible for them to pass droppings. If this blockage is not addressed, the chick is in danger of dying. Pasty butt is so common that you should expect it when you bring home baby chickens, especially if you are new to raising them. They are small and fragile, and a number of stressors can cause it to develop quickly. It is most common in very young chicks, so you will often see it during the first few weeks.

Causes

- Dehydration can make their droppings dry, causing them to stick to the vent.
- Temperatures that are too hot or too cold can cause stress.
- Stress in the form of overcrowding, temperature fluctuations, traveling, and lack of food.
- Poor nutrition.
- Bacterial or viral illnesses.

Symptoms

- Fecal buildup on the vent area.
- Loud chirping.
- Weakness, lethargy, straining, and standing alone.
- Swollen, irritated vent.

Diagnosis

Pasty butt does not require a veterinarian's diagnosis. It is easy to spot and fairly easy to treat on your own. Any dried droppings on the back of your chick are considered pasty butt and need treatment.

Treatment

1. Start by running warm water from the sink in a light stream. Hold your chick in a warm hand, with their back end facing the stream. Let the water run lightly over the chick, ensuring only the vent gets wet. Chicks can become chilled easily, so be sure to keep the rest of their body dry.

2. Using a cotton swab, gently remove the dried droppings from the chick's back. The droppings can dry so much that they pull at the skin. Do this slowly so you don't rip the skin or feathers. If needed, place the chick's backside under water again. You can also use a bit of nontoxic dish soap to loosen the droppings. I find that using my fingers to remove the droppings gently is sometimes easier than using a cotton swab.
3. Once the chick is cleaned off, dry it thoroughly using a paper towel. Keep in under heat until it is completely dry. I have used a heat lamp, a heating pad, or even a heated blanket to warm it back up.
4. Apply a tiny bit of petroleum jelly to the area to help prevent further buildup.
5. Chicks who are especially weak may need a boost with electrolytes and raw egg yolk. Probiotics in their water can also help them recover.

Prevention

- Check chicks daily for pasty butt.
- Keep the brooder away from drafts.
- Give the chicks a heated area as well as a space to get away from the heat.
- Feed appropriate chick feed and provide plenty of water.

Here is an example of what a chick vent should look like. As you can see, there are no droppings on her rear and she is nice and clean!

FAILURE TO THRIVE

Chicks who have their eyes closed while the others are active are ones you need to watch closely.

"FAILURE TO THRIVE" is the term used to describe a chick that is not developing or growing as expected. It may appear less active and grow less than others. You may notice the chick stumbling more and being pushed around in the brooder as it has trouble keeping up. If left untreated, failure to thrive can be detrimental to your chick.

Causes

- Using feed that isn't formulated for chicks, leading to growth and mobility issues.
- Dehydration.
- Genetics or incubation issues.
- Stress caused by extreme temperatures, rough handling, house pets, or poor ventilation.
- Parasites, coccidiosis, respiratory illness, or viral infections.

Symptoms

- Poor development with lack of growth or stunted features.
- Patchy or incomplete feather production.
- Lethargy, unsteadiness, and acting sleepy.
- Refusal to eat or drink.
- Digestive issues such as lack of droppings, constipation, or diarrhea.

Diagnosis

Comparing chicks to others in their brooder of the same age makes diagnosis easier. There is no single diagnosis, as failure to thrive can result from various factors.

Treatment

1. Check that your chick is eating proper food, drinking water, and that the brooder is the correct temperature. If the chick is chilled, be sure to warm them up before offering food or liquid.
2. Give the chick raw egg yolk a few times a day. If they will not eat on their own, use an eyedropper. Place one drop at a time in the side of their beak so they don't aspirate.
3. Add electrolytes to the water to help chicks gain energy and encourage eating and drinking.
4. If your chick still fails to thrive despite improvements in diet, hydration, and environment, consult a veterinarian. They can help identify underlying health issues or recommend specific treatments. Humane euthanasia may be the best option for a chick who continues to fail.

Prevention

- Put chicks under the heat immediately after arrival.
- Keep very young chicks indoors where temperatures don't fluctuate.
- Show each chick how to eat and drink, and observe to make sure they are successful.
- Handle new chicks gently.
- Wash your hands before picking them up.

MUSHY CHICK DISEASE

A closed abdomen looks similar to a belly button while an open abdomen could have a protruding yolk sac.

MUSHY CHICK DISEASE is also known as omphalitis, navel ill, or yolk sac infection. It is a bacterial infection that sets in around the chick's navel and yolk sac. Before a chick hatches from an egg, the last thing they do is absorb the yolk sac into their abdomen. The abdomen closes and forms the navel. Sometimes, chicks hatch before that navel has fully closed, and bacteria can be introduced into the yolk sac or surrounding area. If a chick hatches in unsanitary conditions, such as a soiled nest box or an incubator that was not properly sanitized, it can be one of the most common causes of death in the first week of life.

Causes

- Unhealed or improperly closed navel.
- Contaminated incubation environment.
- Unsanitary egg handling or incubation of soiled eggs.
- Poor brooder hygiene.
- Assisted hatch or premature removal of the chick from the shell.

Symptoms

- Lethargy, reluctance to move and eat.
- Swollen, moist, red navel.
- Distended abdomen.
- Failure to thrive.
- Refusal to leave the heat source.
- Signs of infection, such as foul smell or discharge, from the navel.
- Mortality within the first week, give or take.

Diagnosis

Clinical signs are usually enough to determine the cause of death. A visibly infected navel, an unabsorbed yolk sac, and swelling are easy to spot. While most wouldn't proceed with a necropsy, it would reveal inflammation and cellulitis in the yolk sac.

Treatment

There is no effective cure for mushy chick disease. By the time a vet could assess the infection, run a culture, and provide antibiotics, the infection is too advanced and the chick usually dies. Other than supportive care, quality of life should be assessed to determine whether euthanasia is appropriate.

Prevention

- Only set clean eggs in an incubator.
- Use a sterile incubator and clean it before and after each use.
- Maintain recommended incubator temperature and humidity levels.
- Keep brooders sanitary.
- Avoid helping chicks out of the shell too early, if at all.
- Wash your hands before handling newly hatched chicks.

SPLAY LEG

This chick has classic splay leg. She is unable to pull her legs underneath her body.

SPLAY LEG, OR SPRADDLE LEG, is a condition in which one or both of a chick's legs are positioned out to the sides instead of under the body. Looking at the chick, you would assume they are in gymnastics class, practicing their splits! When a chick hatches with this condition, it makes it nearly impossible for the chick to walk around and access food and water. Splay leg is easier to correct right after hatching. If you wait too long, the legs may be harder to correct, which means it must be dealt with as soon as possible.

Causes

- Poor nutrition of the breeding stock.
- Incubation issues, such as long periods of incorrect humidity and temperature fluctuations.
- Slippery flooring in the incubator or brooder.
- Extensive use of leg muscles trying to push out of the egg.

Symptoms

- Limited movement, shuffling, or not being able to get around.
- Bruising, injury, or death from the others trampling and bumping into the chick.
- The chick may become permanently deformed.
- Skin irritation from prolonged dragging or shuffling.
- Starvation and dehydration from lack of access to food and water.

Diagnosis

While it is easy to spot a chick with splay leg, it can also be confused with a slipped tendon or even a natural deformation. When you hold up a chick with splay leg, their legs should dangle in the correct position. If the toes are completely turned outward or backward, you may be dealing with something else. You should be able to position their legs into the correct alignment with ease.

Treatment

1. Assess the chick's legs to be sure you are dealing with splay leg. If the legs are easily corrected by hand, you should be able to fix the legs. A second person is helpful in this treatment. Start by having one person hold the chick with one hand and position the legs correctly with the other hand. Cut a piece of self-sticking vet wrap ¼ inch wide by 3 inches long (6 mm by 7.5 cm).
2. Using the strip of self-sticking wound wrap, wrap one side around the base of the legs below the knee and fold it over to secure it in place.
3. Bring the legs together in the correct position and repeat on the other leg. The chick should be able to stand if you have done this correctly.
4. Leave the legs wrapped for 24 hours before removing the wrap and assessing the alignment. I have never needed to leave a chick wrapped longer than 48 hours. During this time, make sure the chick can access food, water, and heat. If the chick is weak and being pushed around by other chicks, give them a separate brooder space during this time.

Prevention

- Monitor humidity and temperature in the incubator.
- Use puppy pads on the brooder floor for a few days.
- Use an incubator with a rubber flooring so the chicks don't slip around when walking.
- If your incubator doesn't come with a rubber bottom, cut a piece of cabinet liner to fit and place it on the bottom during the lockdown phase of incubation.
- Ensure parent stock has adequate nutrition.

BACTERIAL INFECTIONS IN BROODER CHICKS

Keeping a brooder clean will help your chicks develop strong immune systems.

BACTERIAL INFECTIONS in young chicks are a major cause of illness and mortality during the first few weeks of life. These infections can occur when chicks are exposed to pathogenic bacteria in the brooder, from the egg or the mother, or as a result of other stressors such as respiratory irritation or poor hygiene. Common bacterial pathogens can affect the gut, respiratory system, eggshell, or internal organs.

Causes

1. COLIBACILLOSIS

Colibacillosis is caused by the avian pathogenic *E. coli*, which is a bacterium that causes disease when a chick's immune system is compromised or exposed to heavy environmental loads. It is spread through contaminated bedding, water, feed, and droppings.

Symptoms

- Weak, listless chicks.
- Poor weight gain.
- Gasping, rapid breathing.
- Swollen or infected yolk sac in newly hatched chicks.
- Necropsy may show signs of septicemia or organ failure.

2. PULLORUM DISEASE

Pullorum is caused by *Salmonella pullorum*, a bacterium passed through the egg or from contaminated equipment, surfaces, feed, or water or infected flock members.

Symptoms

- Weakness and lethargy.
- Huddling near a heat source.
- White diarrhea or pasty butt.
- High mortality in chicks under 4 weeks old.

3. MYCOPLASMOSIS

Mycoplasmosis can occur when a small bacterium (*Mycoplasma gallisepticum*) causes chronic respiratory disease. It is spread by direct contact or vertical transmission.

Symptoms

- Sneezing, nasal discharge, and coughing.
- Reduced feed intake and slow growth.
- Morbidity is high, especially if a secondary infection occurs.

4. INFECTIOUS CORYZA

A bacterium called *Avibacterium paragallinarum* causes an acute upper respiratory infection in chicks and chickens.

Symptoms

- Nasal discharge, sneezing.
- Facial swelling.
- Reduced activity and appetite.

Diagnosis

Diagnosis of bacterial infections typically involves assessing clinical symptoms and risk factors, including brooder cleanliness, biosecurity, and the chick's origin. Laboratory tests, such as bacterial culture, PCR, or necropsy findings, may be used to confirm specific pathogens.

Treatment

1. Start with supportive care such as clean water, high-quality nutrition, proper brooder temperature, humidity control, and isolating affected chicks to prevent the spread.
2. Antibiotics should be selected based on culture results and veterinarian recommendations.

Prevention

- Good brooder hygiene and proper ventilation.
- Biosecurity.
- Proper nutrition and clean water.
- Isolation of new birds.

ASK THE POULTRY DOC

ANTIBIOTICS AND VETERINARY CARE

Federal regulations now require veterinary oversight of all medically important antibiotics used in backyard poultry to combat the global threat of antimicrobial resistance (AMR). Antibiotics are not "cure-all" powders; they are targeted tools effective only against specific bacterial infections and do nothing for viruses or parasites. Using them without a veterinary diagnosis risks promoting the development of resistant superbugs. Veterinary oversight is essential for calculating proper dosing and establishing mandatory withdrawal periods to ensure your eggs and meat are safe for human consumption and free of illegal drug residues. Using antibiotics only when necessary and exactly as prescribed preserves their effectiveness for the future.

VIRAL INFECTIONS IN BROODER CHICKS

Healthy chicks look different from sick chicks. Notice the clear eyes and alert posture of these chicks.

VIRAL INFECTIONS ARE ANOTHER common disease category in chicks and pullets. Viruses can spread rapidly through direct contact, contaminated equipment, or feces. They often damage organs or immune tissues. Because many viral infections lack a specific cure, prevention and supportive care are essential.

Causes

1. INFECTIOUS BRONCHITIS VIRUS

Infectious bronchitis virus is highly contagious and spreads rapidly in flocks via droplets, dust, equipment, or poor biosecurity. Breeding stock can be vaccinated to pass protection to the offspring.

Symptoms

- Nasal discharge and watery eyes.
- Sneezing, coughing, labored breathing.
- Reduced feed intake and weight loss.
- Permanent respiratory damage and poor growth are possible.
- Necropsy may show mucus accumulation and inflamed airways.

2. INFECTIOUS BURSAL DISEASE (Gumboro Disease)

This virus infects and damages the bursa of Fabricius, which is an organ key to immune development in young chickens.

Symptoms

- Depression, ruffled feathers.
- Watery diarrhea.
- Dehydration.

3. MAREK'S DISEASE VIRUS

This is a herpesvirus that is spread easily through dust, dander, and the environment. Most hatcheries will vaccinate for Marek's disease. It is very common and hard to treat, so vaccination is suggested.

Symptoms

- Limb paralysis, general weakness.
- Stumbling, unsteady gait.
- Weight loss, poor growth.
- A necropsy may reveal nerve enlargement and lymphoid tumors.

4. NEWCASTLE DISEASE VIRUS

Newcastle is a paramyxovirus that spreads through direct contact, respiratory secretions, and contaminated materials. Biosecurity is highly suggested, and vaccines are available where appropriate.

Symptoms

- Respiratory distress, gasping, nasal discharge.
- Tremors, twisted neck, paralysis.
- Sudden death with severe strains.

5. AVIAN ENCEPHALOMYELITIS

This virus spreads through the fecal-oral route and, in some cases, from infected breeders to their eggs. Breeding stock may be vaccinated to protect chicks through maternal antibodies. It is most common in chicks under 4 weeks old.

Symptoms

- Neurological signs, tremors, unsteady gait.
- Weakness and poor growth.

Diagnosis

Flock history and clinical presentation are both used to initiate the diagnosis. Veterinarians can use tests such as PCR, viral isolation, serology, or necropsy to determine which virus is present. Some online retailers sell PCR tests that can be completed at home.

Treatment

1. Early detection and supportive care are important because there aren't any medications that can treat viruses. A clean, dry brooder, fresh water, good nutrition, and limited stress will help support sick chicks.
2. Beware of secondary infections that can take hold when the immune system is weakened.

Prevention

- Strict biosecurity.
- Proper brooder maintenance.
- Vaccination for chicks or parent stock, where appropriate.

ASK THE POULTRY DOC

ANTIBIOTICS AND VIRAL PROTECTION

Because viral infections generally lack curative medications—leaving you with only supportive care options like hydration and heat—vaccination is the cornerstone of a successful preventive medicine strategy. It acts as a proactive shield against highly contagious threats that can cause permanent damage, such as the immune suppression caused by infectious bursal disease or the lymphoid tumors associated with Marek's disease.

A robust vaccination protocol protects your flock through two distinct mechanisms:

1. **Maternal Immunity:** Vaccinating breeding stock allows antibodies to be passed to the egg, protecting young chicks from infectious bronchitis and avian encephalomyelitis (which affects chicks under 4 weeks old) during their most vulnerable stages.
2. **Direct Protection:** Administering vaccines to chicks, often at the hatchery level, is critical for pervasive environmental hazards like Newcastle disease and Marek's, where biosecurity alone may not be sufficient to stop transmission.

By investing in vaccination, you shift your management style from reacting to severe clinical signs, such as paralysis, tremors, and respiratory distress, to preventing them entirely.

CHICK TIP While some bacterial infections, such as Salmonella, *E. coli*, and Campylobacter, can be transmitted to humans, the viral infections listed here are not known to infect humans. It is recommended to wash your hands before and after handling chicks.

FUNGAL INFECTIONS IN BROODER CHICKS

While all of these infections seem scary, most chicks will never experience them.

FUNGAL INFECTIONS in poultry are caused by environmental fungi or yeasts that invade tissues, especially when chicks are exposed to large numbers of spores or when their immune systems are weakened by illness, stress, or poor brooder maintenance. Fungal infections are less common in brooder chicks than bacterial or viral infections.

Causes

1. ASPERGILLOSIS (BROODER PNEUMONIA)

Brooder pneumonia is caused by inhaling spores of the Aspergillus fungus. The spores are abundant in the environment and thrive in moldy bedding, damp litter, contaminated feed, and dusty brooders. Young chicks or those that are immunocompromised are more susceptible.

Symptoms

- Labored breathing, silent gasping.
- Rapid respiratory rate.
- Weakness, lethargy.
- Stunted growth.
- Sudden death.

2. CANDIDIASIS (THRUSH/YEAST INFECTION)

Candidiasis is caused by the yeast *Candida albicans* that overgrows in the digestive tract. It is typically caused by prolonged antibiotic use, poor hygiene, vitamin deficiencies, parasite load, or stress.

Symptoms

- Thickened white patches in the crop, mouth, or throat.
- Poor appetite, weight loss.
- Lethargy, listlessness.
- Lesions in the lower intestines.

HATCHING ISSUES

A magnificent look into the shell. It's hard to believe there is enough room in an eggshell for a chick to develop!

HATCHING CHICKEN EGGS can be a rewarding yet delicate process. Even under ideal conditions, a variety of problems may arise during incubation and hatching. Environmental factors such as temperature, humidity, ventilation, and egg handling all play a role, and abnormalities in chick position or development can further complicate the hatch. Understanding these common issues helps you give each chick the best chance of thriving from shell to coop. I have personally hatched hundreds of chicks in my at-home and classroom incubators and have encountered many issues I never thought I would face. Always be prepared for chicks that die inside the shell or shortly after hatching.

Common Hatching Problems

1. **Shrink Wrapping:** The chick pips the shell, but the membrane dries and adheres tightly to the chick, preventing movement and, therefore, hatching. It is often caused by low humidity, opening the incubator too soon, or poor positioning.

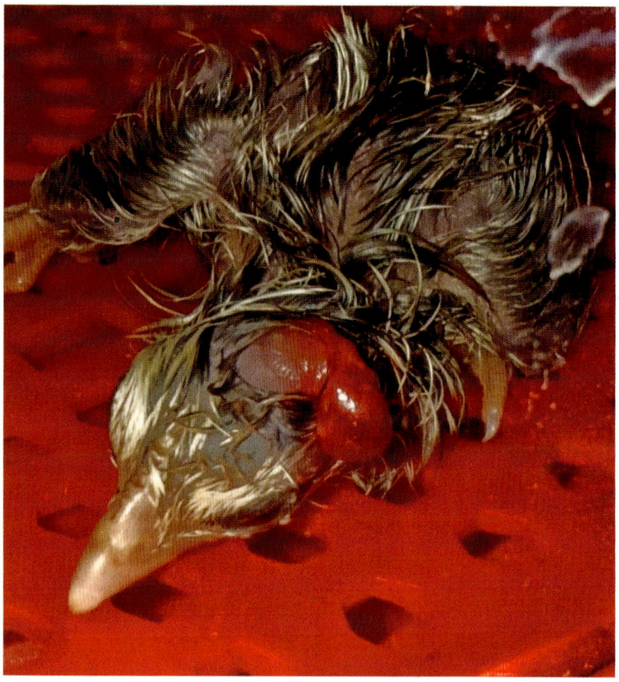

2. **Dead Embryos During Incubation:** This may occur at any stage of development. It can be caused by poor storage, rough handling, incorrect humidity during incubation, and temperature fluctuations during shipping or incubation. You may detect the issue by candling the egg. Embryos may die later in incubation due to incorrect incubation conditions, rough handling, infection, or genetic issues. The most common cause is high humidity, which can drown the chick in the shell. This embryo died early, as indicated by a blood line across the egg rather than veining.

3. **Deformed Chicks:** This can happen due to high incubation temperatures, improper turning, or nutritional deficiencies in the breeding stock. Deformation can appear anywhere on the body, often in the feet or head. This chick had an underdeveloped skull and perished soon after hatching.

CONCLUSION

CARING FOR YOUNG CHICKS is all about creating a healthy start. Once you raise a few clutches of chicks, you'll be a pro at brooder cleaning, nutrition, and common ailments. Practice makes perfect, but hopefully this chapter has helped build your confidence in raising chicks.

As your chicks mature, their social world becomes as important as their physical health. Understanding how they interact, establish a pecking order, and form a balanced flock will help you guide them into adulthood with fewer conflicts and a calmer, happier coop. In the next chapter, we'll explore flock dynamics, including how they communicate, lead, follow, and live together as a group. This topic is one of my favorites!

A happy, healthy chick ready for some fun

This baby sticks close to her mom so she's protected from the others.

9

Flock Dynamics: SOCIAL STRUCTURES, INTEGRATION, AND CARE

Chickens are wonderfully social animals. They rely on one another for safety, resources, social learning, and emotional bonds. Understanding how they relate to one another is one of the most important keys to successful chicken keeping. In this chapter, we'll explore how chickens form bonds, establish order, and communicate. We'll cover everything from how a broody hen nurtures her chicks to how new birds are integrated, how conflicts are resolved, and how to care for your flock through challenges like aggression and loss. By learning to read and support the social world of your birds, you'll strengthen flock harmony, reduce stress, and set the stage for a healthy, resilient group.

BROODY HENS AND THEIR CHICKS

WHEN A HEN GOES BROODY, something deep and instinctive takes over. During this time, she becomes a caregiver, protector, and teacher. A broody hen doesn't just sit on and hatch eggs! She forms a bond that profoundly shapes her chicks' lives. Understanding this relationship offers insight into natural maternal behavior and how best to support it. In my opinion, watching a mother hen with her chicks is one of the best parts of keeping chickens!

What It Means to Be Broody

A broody hen is one that wants to incubate eggs and raise the chicks that hatch. Not all hens become broody in their lifetimes. In fact, most of my hens have never been broody. A few, however, tend to go broody all year long! Some breeds, such as Silkies, Orpingtons, Cochins, and many mixed breeds, are more likely to go broody. Signs include:

- Spending all day on the nest, only breaking to eat and drink.
- Fluffed feathers that rise up when you approach the nest.
- Clucking softly or screeching when disturbed.
- Staying in the nesting position when placed on the ground. (I call them pancake chickens!)

Sarah and her friends won't leave the nests, so they had to stay in the "Henitentiary."

Options for a Broody Hen

In nature, a hen will sit on fertile eggs until they hatch, then raise the chicks and integrate them into the flock. She will remain broody until the chicks hatch. Since your hen isn't living in the wild, you may or may not want her to hatch chicks. As a chicken keeper, you have a few options to help her out.

BROODY BREAK HER

Broody breaking is the process of helping her snap out of her broodiness. I prefer a crate with a wire bottom, propped up on bricks. I give her a roost, food, and water and leave her in the crate for 3 days. This is usually enough to pull her out of the broody trance! I do this when I don't want chicks added to my flock, because allowing her to stay broody week after week without the chance of chicks is not only cruel but also depletes her health.

GIVE HER FERTILE EGGS

If you have a rooster mating your hens, you can give her those eggs to hatch. If you don't have a rooster or want a specific breed, give her fertile eggs from another farm or hatchery. Mark the eggs so you know which to leave with her and which to collect. Check each egg to make sure she didn't steal another egg from the flock. Check on her multiple times a day to make sure she is still sitting on the correct eggs, or separate her and her nest of eggs so she can't hop to another nest. In 21 days, she will hopefully have chicks!

GIVE HER DAY-OLD CHICKS TO ADOPT

Broody hens will sometimes adopt chicks you or a hatchery has hatched. I believe it works best once a hen has been broody for a few weeks. After dark, sneak each chick under your broody hen and carefully collect the eggs she was sitting on. Leave the hen and chicks alone all night. Hopefully, by morning, she will have accepted them as her own. If she is softly clucking to the chicks, scratching around and showing them food, and squatting so they can get warm under her feathers, you have a successful mother hen!

Supporting a Mother Hen

When you have a mother hen in the flock, it is your job to support her instincts and keep her and the chicks safe and comfortable. Even though she will likely protect the chicks from the flock and other dangers, you can still make that easier on her, and you should observe to make sure she is successful.

Use an egg-safe ink stamp or a wax crayon to mark the shell, not a marker with ink that may move through the pores.

Brenda couldn't be happier to have her own babies to raise, and she didn't even care that I bought them.

Provide her with a safe space to get to know the chicks. I use the integration station under my shelf. The chicks and the mother hen can bond while still allowing the flock to check them out and get used to their presence. The mother and chicks should eat food safe for the chicks, and they will need safe water. I like to leave them in this space for about 5 days. By then, they are following their mother's lead and are a bit quicker and stronger, able to navigate the rest of the coop.

After a few days, you can try letting her and the chicks join the flock. It is important to let them out on a day when you are free to observe their interactions. A good mother hen will protect the chicks by directing them to safe spaces or putting the flock in its place. You may see her fight off other hens or pick fights for no reason. She is telling the rest of the flock that she and her babies should not be messed with. This is a good thing! Let her handle the flock as long as she isn't drawing blood. A few squabbles and the flock will leave the babies alone. The flock should all be eating feed safe for the chicks until they are old enough to move up. Offer crushed oyster shells on the side for the hens who need calcium.

In my experience, most mother hens will raise their chicks for about 6 weeks. I have a hen who gives up on week 3 every single time. She acts as if she has never seen those chicks before, and she is DONE! I have another who raised a chick all the way to 21 weeks, when the chick laid her first egg! Luckily, by the time they give up, the chicks should be integrated enough into the flock that you won't have to step in.

SEPARATE IS BEST

Unfortunately, I have witnessed horrible incidents in which a hen in the flock attacks eggs or chicks. I have also seen broody hens leave their nest of eggs to eat and return to the wrong nest, leaving the partially developed chicks to perish. When allowing a hen to hatch chicks, a separate space might be best.

FLOCK HIERARCHY AND PECKING ORDER

Jenny and Rosalina wait their turn for some grubs (they are two hens near the top of the pecking order).

ONE OF THE MOST IMPORTANT and fascinating aspects of chicken social life is the pecking order. This system establishes and maintains a social hierarchy within the flock. It isn't just about who pecks whom; it's a structured ranking that shapes everyday life in the coop and helps reduce constant conflict over resources. The term "pecking order" originated from early observations that dominant birds literally pecked those below them in rank. Chickens at the top of the pecking order have priority access to food, water, preferred roosting spots, and mates. Those in lower-ranking positions typically wait their turn to avoid confrontation. In this dominance hierarchy, each bird has a rank relative to every other bird in the flock.

How Pecking Order Forms

Dominance relationships develop as chickens interact. Chicks begin establishing social rank as early as a few days old. You may notice pecking, chasing, posturing, and sparring, all of which help determine who comes out on top. While changes can occur, this order usually holds through adulthood.

Initially, these squabbles are innocent and playful, but as they mature, interactions become more ritualized. Birds may use body language or engage in full-on fights to assert dominance. Chickens may avoid the top hen or pick on one or two bottom hens. Once a stable hierarchy is established, aggressive interactions usually decrease. However, changes in flock dynamics, such as adding new birds or losing a member, or changes in the environment or resources, can challenge the hierarchy and reestablish order. The pecking order can be seen in these daily experiences:

- **Eating and drinking:** Dominant birds often eat and drink first, while lower-ranking birds wait until they are finished. Providing multiple feeders helps reduce fighting and ensures everyone can eat.
- **Roosting and nesting:** Higher-ranking chickens typically claim their preferred sleeping spot or favorite nest box. If a lower-ranking bird is in their spot, you will see them peck the other chicken until they move.
- **Mating:** In flocks that contain a rooster, dominant males often take the lead in mating behaviors. Lead hens are often mated more frequently. If there is no male, lead hens may perform mating behavior on other hens.
- **General behavior:** Hens lower in the pecking order may dodge others or avoid eye contact.

Fudge is the head hen and very clearly has authority over the feeders and nest boxes. She even inserts herself as a mother figure to babies belonging to other hens!

Normal vs. Problematic Behavior

The pecking order hierarchy helps reduce conflict and provide structure. As a result, some pecking and squabbling is entirely normal, even in a healthy flock. However, there is a difference between everyday social behavior and patterns that indicate stress, bullying, or serious welfare issues.

TYPICAL PECKING ORDER RANKS

1. Top Dominant Bird (Alpha)
   - Head of the flock
   - Can be male or female
   - First to access resources
   - Others often move out of their way

2. Second-Rank Bird (Beta)
   - Supports the alpha's dominance
   - May dispute minor claims
   - More assertive than the rest

3. Middle-Rank Birds
   - A group that takes turns with access to resources
   - Less dominant
   - Often stable and peaceful with each other
   - Submit to the higher-ranking birds
   - Sometimes assertive over lower birds

4. Subordinate Birds
   - Yield readily to higher ranks
   - Eat, drink, and roost after the others
   - Avoid conflict and dodge away

5. Lower-Rank Bird (Omega)
   - The most submissive
   - The most picked on or bullied
   - Last to get resources

NORMAL SOCIAL BEHAVIORS

- **Subtle pecking during rank establishment:** Chickens may peck or chase briefly as they sort out who leads and who follows. These interactions don't cause harm or break skin, and should taper off as the order is established.
- **Priority pecks around resources:** Pecking that occurs once or twice to assert dominance around feeders, waterers, or scraps is normal behavior.
- **Nonviolent displays of dominance:** Wing flapping, feather fluffing, brief chasing, and posturing are normal social signals.

PROBLEMATIC SOCIAL BEHAVIORS

- **Persistent targeting of one chicken:** Ongoing harassment of a chicken who is inactive or in a safe space goes beyond the occasional reminder of rank.
- **Feather pecking:** Mild feather pecking or preening is typical behavior, but repeated pecking that removes feathers or injures skin isn't.
- **Physical aggression with wounds:** Any fights that draw blood, damage eyes, or cause open wounds are considered aggressive.
- **Isolation and fear:** If a lower-ranking bird is isolated from the flock, unable to obtain food and water, or becomes depressed, this is a response to aggressive bullying.

How to Stop Bullying

When problematic behaviors arise, it is best to intervene to prevent further bullying, severe injuries, extreme feather loss, starvation, or cannibalism. If you feel like your flock's behavior is escalating rather than resolving, it may be time to step in. Here are a few adjustments you can make for your flock:

- Provide multiple feeders and waterers so low-ranking birds can eat without confrontation.
- Ensure adequate space in the coop and run.
- Provide roosting bars, ladders, logs, or climbing spaces so birds at the bottom of the pecking order can get off the floor and out of the way.
- Provide enrichment to the flock, such as treat blocks, hanging vegetables, treat dispensers, and free-range time.

REHOMING ROOSTERS

Rehoming a rooster can be difficult, especially because most small backyard flocks can only accommodate one. Here are my suggestions on where to start:

- Join local backyard chicken-keeping Facebook groups. Many groups allow you to list your roosters for rehoming. Consider offering him a hen.
- Use homesteading apps that let you list what you have for sale.
- Utilize local farms that will process roosters for meat (not ideal if you have become attached to him, but at least his life was purposeful).
- Contact local 4-H, FFA, or other youth groups, as they often need well-handled roosters.
- Attend local farm swaps or animal auctions.
- Reach out to rescue groups or sanctuaries.
- Donate him for processing, pay a farm to process him for you, or learn to process meat yourself.

- Monitor your flock's health, as a chicken's illness or weakness can trigger bullying.
- If all else fails, remove the bully for a few days. Give them a safe space with food and water, but far from the flock, so when you bring them back, the pecking order may adjust appropriately.

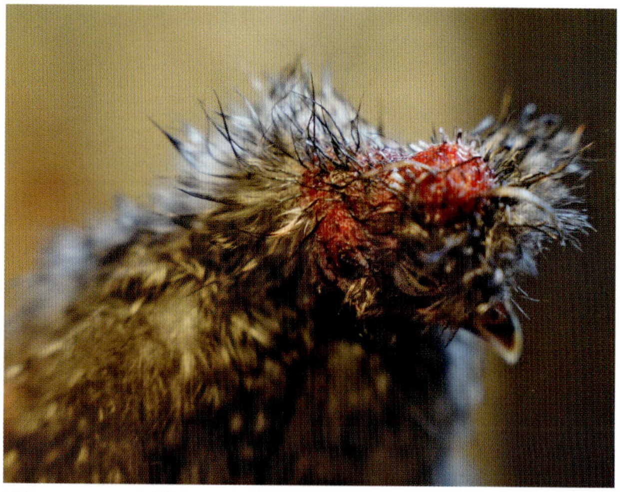

This poor, sweet Silkie was attacked by the flock members a few days after what I thought was a successful integration.

Big Mike has been an amazing rooster for our flock. He is very diligent at protecting the hens from hawks, dogs, and foxes.

A Rooster's Role in the Flock

A rooster isn't just the protector and the morning wake-up call. He also serves distinct social and practical roles. In a flock with a rooster, his presence shapes hierarchy, safety, behavior, and social organization. When there are two or more roosters in a flock with hens, they establish a pecking order among themselves. To minimize fighting between roosters, maintain a hen-to-rooster ratio of at least 10:1. However, I prefer a higher hen-to-rooster ratio. Key natural roles of a rooster include:

- **Flock protection:** A rooster watches for predators, alerts the flock, and puts himself between the danger and the hens. I have seen my rooster risk his life for his flock many times.
- **Territory defense:** His crow helps establish territory and discourage rivals.
- **Social leadership:** The rooster is typically above the hens in the hierarchy, but he normally doesn't need to peck them into place. Roosters command respect just by their presence and mating behavior.
- **Flock peacekeeper:** Roosters will break up fights between hens. My rooster will run across the yard to stop a hen fight! He gives them each a peck on the head or puffs up his chest and they quickly move on.
- **Breeding:** The rooster will mate with the hens, which fertilizes the eggs so they can grow into chicks if kept warm.

ROOSTER AGGRESSION TOWARD HUMANS

Roosters may exhibit behavior that can be misinterpreted as aggression toward people. When they perceive your movements or actions as a threat to their flock or feel their dominance is being challenged, they may chase, posture, or kick. Most roosters go through a hormonal phase before fully maturing. During this phase, they tend to assert their dominance more often. Here's how to earn your rooster's respect:

- Stand your ground and stay calm.
- Don't run away, scream, or throw things in his direction that will intensify his aggression.
- Don't turn your back on an aggressive rooster or outpace him when walking nearby.
- Avoid chasing and grabbing his hens in his presence.
- Feed him from your hands as often as you can.
- If he'll let you, calmly pick him up and gently stroke his feathers.
- If all else fails, provide him with a fenced area to keep the humans safe or rehome him.

HOW TO TRIM A ROOSTER'S SPURS

1. Secure the rooster with a partner. It is best to take him off the roost at night rather than chase him. Hold him securely in a towel and avoid letting his beak get near your face.
2. Make sure his spurs are clean by wiping them with an alcohol wipe.
3. Use sharp dog nail clippers, side cutters, or a rotary tool with a grinding bit to trim the sharp end of his spurs. Avoid cutting too far, or you'll nick the central blood vessel and cause bleeding. Trim small amounts at a time. Apply pressure and styptic powder or cornstarch as needed to stop bleeding.
4. File the ends smooth so they're not sharp or jagged.

INTEGRATION OF NEW FLOCK MEMBERS

You would never know that some of these birds were once new to the flock and ostracized by the others.

INTRODUCING NEW BIRDS or returning separated chickens to the flock is one of the more delicate yet important parts of chicken keeping. Chickens form strong social bonds and clear hierarchies, so changes can cause stress, competition, and confusion if not managed thoughtfully. Whether you are welcoming new chicks, introducing older pullets, or returning an injured hen to the flock, the key is to balance patience, structure, and observation. Understanding how to make these transitions as smooth and low-stress as possible will help establish a flock that lives peacefully and accepts newcomers with ease. Done well, integration strengthens your flock's resilience, deepens social bonds, and sets the stage for a calmer, more cohesive group.

Integrating Chicks into the Flock

Introducing chicks to an established flock can be tricky if the right steps aren't taken. Chicks are small, vulnerable, and socially different from adults. The goal is to set them up for success so the flock doesn't attack or bully them and they become accepted members of the flock. The flock will instinctually sort itself into a pecking order. Without proper preparation, they can become targets of curiosity, pecking, or even aggression. While some pecking and rank assertion is normal, excessive harassment can lead to injury in chicks who are less able to cope.

TIMING

- Wait until the chicks are fully feathered so you don't have to worry as much about temperatures. This is usually around 6 to 8 weeks.
- Some chicks do better when they are introduced young; that way, they take their place in the pecking order without fighting back. If you wait until they are close to the size of the flock members, they may try to assert themselves, making integration more difficult.
- Choose a week when you are available to visit the coop many times a day.

PROCESS

1. Prepare an integration station for the newbies to hang out while the flock gets to know them safely. This could be a crate, a temporary partition, or a makeshift fencing structure within the run. Make sure the flock can see the chicks and get close to the enclosure. You don't want the chicks to fit through the bars or wire, and they should have enough space inside so that if they move to the middle, the hens can't peck them through the openings. At night, the chicks can go into a crate in the coop or back to their brooder.

2. After a week in the integration station, it's time for short supervised interactions. It's best to let them interact in a neutral location. I like to let the flock out to free-range, then let the chicks out to explore the run while the flock is distracted. You can distract the flock with treats or food scraps during these short, supervised periods.

3. If these sessions are going well, you can allow them more time together. One trick I use is to prop the door open so the chicks can go in and out of their integration station, but the bigger chickens can't fit inside. This gives them a safe zone while still allowing them to integrate into the flock when they feel ready. If the sessions aren't going well, give them more time in the station before trying again.

4. When you feel the chicks are mostly integrated, you can remove the station and slip them onto the roost after dark. Hopefully, in the morning, the flock will be content with the chicks being part of the group. I suggest checking on everything a few times during the day.

The new chicks on the block are safely behind bars, making friends one chirp at a time.

NORMAL INTEGRATION BEHAVIOR

- The hens gently peck at the chicks' feathers or tail, then move on.
- The hens chase the chicks for a few steps to push the chicks into a different direction or away from resources.
- The hens give a hard peck to put the chicks in their place, but then return to normal activity.
- The chicks stick together and avoid the older flock members.
- The hens or roosters vocalize.

PROBLEMATIC INTEGRATION BEHAVIOR

- Hens repeatedly and constantly chase the chicks.
- The hens persistently peck at the same chick.
- The chicks are prevented from accessing food and water.
- The chicks display open wounds.
- Adult birds shake or pick up the chicks.

READY RESOURCES Providing new resources is a great way to make integrations easier!

- Provide additional feeders and waterers in different locations to prevent resource guarding.
- Scatter treats to encourage calm, distracted behavior.
- Add roosts, branches, or hiding spaces to give the chicks a few safe spots.
- Add treat dispensers or other distractions.

Although Blaire would really like to run free with the flock, I know that wouldn't be safe for her. To the integration station, Blaire!

ASK THE POULTRY DOC

PECKING WOUND CARE

In a traumatic pecking wound situation, the immediate priority is prompt wound care to reduce infection risk and stop further damage. Gently clean the area with a diluted chlorhexidine solution to remove debris and lower the bacterial load, then apply a thin layer of sterile, medical-grade Manuka honey to protect the tissue and support healing. Isolate to prevent continued pecking from the flock. Monitor the site closely for signs of infection. The bird should remain separated until the wound has fully healed and feathers begin to regrow; upon reintroduction, observe the flock closely to ensure the area is not targeted again. To prevent future occurrences, consider preventive measures such as pinless peepers (plastic, spectacle-like devices to protect a chicken's eyes from pecking) and increasing available space to reduce stress.

Integrating Pullets or Adult Birds into the Flock

Bringing pullets (teenage hens) or older chickens into an established flock is one of the more sensitive transitions in flock management. Unlike chicks, older birds have a more established pecking order and social identity. Established flock members may view newcomers as competitors or intruders, especially when the newcomers are similar in size. A thoughtful, gradual transition helps minimize aggression and injury.

TIMING

- Pullets or hens can be integrated at any age.
- Quarantine the new arrivals for 4 weeks before integration. Monitor health and consider deworming and external parasite control.
- Choose a time when you are available to manage the integration and when the flock isn't under external stress.

PROCESS

1. Quarantine your new birds to assess their health and to make sure new health issues don't arise. Often, viruses can show signs days or weeks after exposure. This also allows your newcomers to settle after travel. I will usually treat my new members for internal and external parasites in case they have a small, undetected load.
2. Use an integration station or a smaller temporary coop inside or next to the run. This allows them to see each other without hurting one another. Keep them in this area for 1 to 2 weeks, depending on the flock's response. If the flock is being aggressive against the crate, keep the new birds in there longer.
3. Once both groups seem calm around each other, allow short, supervised encounters. These should take place outside the coop and run so the flock is distracted by foraging and away from resources they may guard. Scatter treats if additional distraction is needed. If things go well, allow more time, but stay nearby to observe behavior.
4. Once the birds are familiar with each other, calm, and comfortable navigating the space, you can open the partition and allow them to mingle freely. You may also place them on the roost after dark to reduce the original flock's response. Be sure to provide additional feeders, waterers, and other resources.

NORMAL INTEGRATION BEHAVIORS

- Pecking to reestablish pecking order.
- Brief chasing and pecking that does not result in injury.
- Posturing at each other as part of social sorting.
- New members may stick together and avoid the flock for a few months.

PROBLEMATIC INTEGRATION BEHAVIORS

- Persistent, targeted attacks on any of the new members.
- Any pecking that causes bleeding, open wounds, or a lot of feather loss.
- Continual blocking of food and water.
- Nonstop fighting or lengthy battles.

Reintegrating a Hen into the Flock

When a hen is removed due to illness or injury, her place in the pecking order may change. Chickens quickly adjust their social hierarchy in her absence, and she may return lower in rank. Proper reintroduction can help reduce stress and ease her back into the flock.

CiCi was away for weeks recovering from a beak break. My flock is used to integrations, so I put her on the roost at night, and they accepted her back.

TIMING

- Only bring the hen back once she is physically healed, no longer contagious, and eating on her own. This reduces the risk of injury and of passing viruses or bacteria to the flock.
- Integrate when the flock is calm and free from external stressors.
- The process can take a few days or a few weeks.

PROCESS

1. Start by placing your recovering hen in a secure area where she can see and be seen by the flock and where she can hear them. In this crate or pen, she will need her own food and water. This helps the flock recognize her presence, reducing the risk of harsh interactions.
2. After a few days in the safe integration station, allow her access to the roosts after dark or out of the crate during free-range times. Stay nearby in case intervention is needed.
3. Provide distractions, such as treat dispensers and hiding spots, if the other chickens bother her.
4. Adjust your approach if the other chickens seem to be picking on her. You can give her more time in the integration station or pair her with a calmer buddy.
5. Some flocks will readily accept the hen back if you place her on the roosts with them after dark. Be sure to observe them in the morning to assess their behavior.

NORMAL REINTEGRATION BEHAVIORS

- Brief pecking or posturing.
- Circling or vocal exchange as the chickens assess her identity.
- Brief squabbles that subside quickly.

PROBLEMATIC REINTEGRATION BEHAVIORS

- Targeted aggression or ganging up to intimidate, block resources, or attack.
- Persistent injuries or feather pulling.
- Isolation of the returning hen.

MANAGING LOSS IN THE FLOCK

LOSS IS AN INEVITABLE part of keeping chickens and can be one of the most challenging aspects. Whether it's the quiet passing of an older hen or the heartbreak of a predator attack, loss touches both the emotional heart and the practical day-to-day work of flock life. It's important to recognize normal mortality versus patterns that signal a deeper issue. Finding a deceased bird is tough, but deciding when to cull a bird to respect their quality of life is equally difficult. Supporting your flock and yourself through loss is also something to consider carefully. By approaching loss with sensitivity and practicality, you'll be prepared to care for the situation with confidence and strengthen the overall harmony of the flock.

Understanding Loss

Loss can be an emotionally charged subject for a flock keeper. While the death of a chicken is never easy, it is part of normal backyard chicken keeping. Even in well-managed flocks, some mortality is expected because chickens are living, aging animals. Being outside, they are also subject to a range of health challenges, predators, and accidents. Loss varies by region, management style, breed, and predator events. Many chicken keepers report that a few losses a year in a backyard flock are not unusual.

I don't know if I will ever get over the loss of Louise, my sweet and friendly lap chicken.

NORMAL LOSS

- **Early life:** Young chicks are especially vulnerable to chilling, smothering, viral infections, and poor hatching conditions. Chicks are very fragile, and a small amount of loss in the first few weeks is unfortunately not uncommon.
- **Growth and molt transitions:** Occasionally, a flock member will succumb to stressors such as cold, heat, or heavy feather loss.
- **Age:** Like other animals, older hens decline gradually due to wear and tear on their bodies and reproductive systems, as well as other chronic conditions.
- **Disease and infection:** Viruses, bacteria, and parasitic conditions are major contributors to mortality.
- **Management-related stress:** Poor ventilation, nutritional imbalance, or mishandling can weaken birds and lead to death.
- **Predation and accidents:** Foxes, raccoons, dogs, hawks, and other predators account for many unexpected losses. Natural and accidental incidents can cause death.
- **Behavioral issues:** Stress, boredom, improper integration, or poor flock condition can lead to severe pecking or even cannibalism.

RED-FLAG DEATHS

Although occasional loss is to be expected, certain patterns suggest issues that require closer attention in order to prevent further loss.

- **Multiple deaths in a short period of time:** If death is accompanied by clinical signs such as lethargy, loss of appetite, or respiratory symptoms, it could be a signal that others may die as well.
- **Infectious disease signs:** Sudden death without apparent cause, neurological signs, labored breathing, or unusual discharge should be investigated further.
- **Unusual stressors:** Abrupt weather changes or poor housing conditions can cause death and should be adjusted if they are believed to have caused death in the flock.

Responding to Death

Finding a deceased chicken can be upsetting, and how you respond emotionally and practically sets the tone for both your peace of mind and the health of the rest of your flock. First, take a moment. It's normal to feel sadness when you find an animal you cared for. Their loss matters, and allowing yourself a brief pause before jumping into action is appropriate.

- After assessing the situation, remove the bird from the flock's area using gloves and a clean bag or container to prevent contaminating the environment.
- Look for obvious clues, such as predator marks, scattered feathers, signs of illness, or evidence of trauma.
- Make note of the context. Was the bird alone? Was the rest of the flock acting normal?

After removing the body, observe the flock for a day or two. Watch for changes in appetite, posture, droppings, breathing, and social behavior. Subtle signs can help you determine whether the loss was an isolated event or part of a larger issue. If the others show signs of illness, you may want to send the body for a necropsy or have your vet test the other birds.

Necropsy

A necropsy is the animal equivalent of an autopsy and can provide valuable insights into why a chicken has died. It can give you information that helps protect the rest of your flock or prevent future incidents. Many veterinarians and diagnostic labs offer poultry necropsies at a reasonable price for backyard chicken keepers. What you learn from a necropsy can be invaluable. It can diagnose an underlying disease like Marek's disease, respiratory infections, nutritional deficiencies, and fatal injuries. It allows you to make adjustments to your husbandry and gives peace of mind.

ASK THE POULTRY DOC

NECROPSY

The poultry necropsy process is a comprehensive dissection focused on differentiating normal anatomy and common postmortem changes from actual disease to avoid misdiagnosis. The evaluation begins with an external examination of the skin to distinguish artifacts, like a "green belly" caused by bile imbibition, from pathologies like gangrenous dermatitis. Dissection involves opening the coelomic cavity to evaluate for gross pathology, checking organs, fluid, and air sacs, and noting whether features like fat pads (yellow/greasy) are mistaken for tumors (white/firm) or whether air sacs are cloudy (indicating airsacculitis). The internal exam continues by opening the digestive tract to assess the liver, which should have sharp edges and a smooth surface, and the ceca, which are checked for hard, cheesy cores. The reproductive tract is evaluated, looking for abnormalities such as cauliflower-like masses (ovarian cancer) or an impacted oviduct (salpingitis), while recognizing common variants like the cystic right oviduct. Finally, the immune and structural organs are examined, including the spleen (which should be small and round, not enlarged or mottled), the spongy, fixed lungs, and the bursa of Fabricius, whose size varies significantly based on the bird's age.

My children helped honor Louise by placing flowers over her body.

We like to place a large rock or a perennial bush over the grave to remember special chickens.

HOW TO PROCEED

1. **Preserve the body:** Refrigerate the bird as soon as possible. Do not freeze it, as freezing can compromise diagnostic results.
2. **Contact a vet or diagnostic lab:** Many state animal diagnostic labs and universities accept backyard poultry. Some provide submission forms and specific packaging instructions.
3. **Follow packaging and shipping directions:** Labs often require the bird to be double-bagged, with a submission form sealed to keep it dry. Shipping overnight ensures the specimen arrives in good condition.

Options for Handling a Dead Chicken

1. **Burial**
   - Bury the carcass at least 2 feet (60 cm) deep to discourage scavengers and minimize odor.
   - Avoid burying near wells, streams, or water sources to prevent contamination.
   - Some local ordinances do prohibit backyard burial.
   - Use a large rock, a new plant, or a tree to mark the spot. We love planting hydrangea bushes over the burial spot.
2. **Composting**
   - Build a compost pile using carbon-rich materials, the carcass, poultry litter, and food scraps.
   - Composting can convert a deceased chicken into stable material for your land.
3. **Incineration**
   - Burning the carcass in a controlled setup can destroy pathogens for biosecurity. There is some odor, so be mindful of neighbors.
4. **Disposal**
   - Double-bag the body and dispose of it through your municipal trash pickup if allowed.
5. **Veterinary Cremation**
   - If the chicken was a pet and you want a memorial option, your vet's office may offer cremation services.
   - Local practices vary widely, so check with your vet.

When to Cull

Culling is the humane practice of removing a bird from your flock when their quality of life is compromised or when they pose a risk to the health of others. The term "culling" is also used when removing a bird from a breeding program and rehoming it or selling it to another farm. It's a difficult decision, but part of responsible flock management.

The goal of culling isn't punishment. It is done to prevent suffering or widespread disease. Healthy birds can live several years, but conditions such as severe, chronic illness, irreversible injury, poor quality of life, or suffering indicate that your chicken may need to be humanely euthanized. Signs that may justify culling include:

- Severe, nonresponsive injury (such as severely broken legs that can't heal).
- Chronic disease with no realistic chance of recovery.
- Reproductive conditions that cause severe prolapse or suffering.
- Neurological conditions that impede survival.
- A contagious disease may warrant culling, especially if the bird can't survive the disease and is putting the others at risk.

Getting attached to chickens makes their loss harder. Trix, a three-legged rooster, stole my heart but died during surgery, which devastated me.

ASK THE POULTRY DOC

THE CULLING PROCESS

When considering euthanasia options, cost is a significant factor for many backyard poultry owners. While a veterinary clinic provides a "gold standard" experience with professional sedation and barbiturates to ensure a peaceful end, not all owners can afford the associated costs of a clinic visit. For those who find professional services financially inaccessible, the American Veterinary Medical Association (AVMA) approves a couple of at-home methods.

- **Cervical dislocation (CD):** This manual method is the most commonly used and involves separating the cervical vertebrae at the base of the skull to sever the spinal cord. When performed correctly, it is effective for birds weighing under 6.6 pounds (3 kg) and results in immediate loss of consciousness and rapid death.

- **Mechanical CD:** Tools like the Koechner Euthanizing Device (KED) can remove human error and ensure a clean anatomical break for those concerned about their own strength or precision.

Providing these home-based alternatives ensures that all keepers, regardless of financial circumstances, can uphold their final responsibility of stewardship by providing a humane and dignified death for their birds.

Support Your Flock After Loss

Once a loss has occurred, both the flock's behavior and the environment they live in benefit from deliberate support and care. Chickens notice when a flock mate is gone, and if the loss was traumatic, they can become frazzled. They may vocalize more, seem agitated, or congregate together. These behaviors are normal social responses and often diminish with time. You can support your flock and their environment by doing the following:

- Give the birds a consistent routine.
- Offer distractions like free-range foraging.
- Allow them to reassess their pecking order, especially if the deceased hen was at the top or bottom of the order.
- Clean and disinfect any area where the bird spent significant time.
- Replace bedding if it was soiled.
- Clean feeders and waterers if there is a risk of contamination.
- Monitor the flock for appetite changes, lethargy, labored breathing, and unusual droppings.

Supporting Yourself After Loss

Losing a beloved chicken can be emotionally challenging. Chickens become part of the rhythm of our days, and it's okay to feel sadness or guilt. Supporting yourself is just as important as caring for the rest of the flock.

- Acknowledge your feelings and give yourself permission to feel grief.
- Perform a ritual, like placing a burial marker, picking flowers to place in the grave, or lighting a candle.
- Reach out to friends, other chicken keepers, or an online community that understands your emotional attachment.
- Take the opportunity to learn from the death so you can better serve the rest of your flock.

CONCLUSION

UNDERSTANDING CHICKENS as social beings is at the heart of effective, compassionate flock care. From the complex pecking order that organizes daily life to the carefully managed introductions of new members and the realities of loss, each element of social structure shapes how your flock functions and thrives. Recognizing these patterns enables you to respond with thoughtfulness, transforming challenges into opportunities for healthier birds and more confident care.

Social structures give your flock a framework for predictability and harmony, but they also require gentle stewardship. A successful, productive, social flock is what makes chicken keeping so rewarding. As you move forward, remember that what you learn will carry through to your husbandry, and that every interaction is part of the fulfilling journey of chicken keeping.

FLOCK TRACKER

NAME	BREED	HATCH DATE	PURCHASED FROM	EGG COLOR	FIRST EGG DATE	DATE OF PASSING

FLOCK TRACKER

NAME	BREED	HATCH DATE	PURCHASED FROM	EGG COLOR	FIRST EGG DATE	DATE OF PASSING

FLOCK TRACKER

NAME	BREED	HATCH DATE	PURCHASED FROM	EGG COLOR	FIRST EGG DATE	DATE OF PASSING

CHICKEN PROFILE

NAME

HATCH DATE

BREED

eggs

FIRST EGG LAID ON

EGG COLOR

I CAME FROM

Temperament

fun fact

Date Passed

How

CHICKEN PROFILE

NAME

HATCH DATE

BREED

eggs

FIRST EGG LAID ON

EGG COLOR

Temperament

fun fact

I CAME FROM

Date Passed

How

CHICKEN PROFILE

NAME

HATCH DATE

BREED

eggs

FIRST EGG LAID ON

EGG COLOR

I CAME FROM

Temperament

fun fact

Date Passed

How

CHICKEN PROFILE

NAME

HATCH DATE

BREED

eggs

FIRST EGG LAID ON

EGG COLOR

I CAME FROM

Temperament

fun fact

Date Passed

How

RESOURCES

Purely Chickens Content from Nikki Husted

- **Instagram**
  @purelychickens
  instagram.com/purelychickens
- **TikTok**
  @purelychickens
  tiktok.com/@purelychickens
- **YouTube**
  @purelychickens
  youtube.com/@purelychickens
- **Facebook**
  @purelychickens
  facebook.com/purelychickens
  facebook.com/groups/purelychickens
- **Website**
  purelychickens.com

Web Resources

- **The Poultry Doc (thepoultrydoc.com):** Offers one-on-one veterinarian advice, educational articles and videos, a symptom checker, and diagnostic test kits.
- **PoultryDVM (poultrydvm.com):** Offers a chicken symptom checker, an avian veterinarian list by state, and articles written by a veterinarian for all of your poultry ailment concerns.
- **American Poultry Association (amerpoultryassn.com):** A good way to learn about exhibition poultry.
- **Animal and Plant Health Inspection Service (aphis.usda.gov):** A resource for poultry health, biosecurity, and illness outbreaks.

Favorite Brands and Products

- **Eaton Pet and Pasture**
  eatonpetandpasture.com

  I use this company as a one-stop shop for chicken keeping! Eaton Pet and Pasture is a USA-grown, pesticide-free, and sustainably sourced company. I like the quality of their products and their environmental friendliness! My favorites include:

  - **Hemp bedding:** I use this for chick brooders and on the floor of my chicken coop. It is low-dust and absorbent, so the areas stay dry and clean!
  - **Revitalime:** I use this for pest, odor, and moisture control. I also use it on my shelf to dry up wet spots and keep flies away. I add it to the dust bath area to repel mites and lice.
  - **Roost Refresh:** This is a hemp and herb blend that I use for the shelf under the roosts. It can also be added to the brooder or coop bedding.
  - **Aspen Nest Pads:** These are my favorite nest box bedding for clean eggs!
  - **Organic and non-GMO chicken feed:** Use my code "purelychickens" to save!

- **RentACoop**
  rentacoop.com

  RentACoop is one of my favorite places to find chicken-keeping supplies! I love their feeders, waterers, treat dispensers, incubators, and chick brooder supplies!

- **Better Flock**
  betterflock.com

  I use Better Flock AvaPlex to supplement my flock in times of stress or sickness. I use Nutriplex to boost overall health and wellness.

  Use my discount code "purelychickens10."

Hatcheries

Here are a few hatcheries I have used and trust.

- **Meyer Hatchery: Polk, Ohio**
  Use my discount code "purelychickens."
- **Alchemist Farm: Sebastopol, California**
  A sustainable and ethical farm run by my friend Franchesca and her family! Use my discount code "purely10."
- **My Pet Chicken: Brooks, Georgia**
- **McMurray Hatchery: Webster City, Iowa**
- **Buckaboo Farm: North Tustin, California**
  A women-owned farm and small business for hatching eggs and live birds, run by my friend Tiffany.

More about The Poultry Doc

The Poultry Doc is an online resource built just for backyard flock owners, bringing together science-backed educational resources, a curated e-commerce marketplace, flock dashboards, and a veterinarian-led community forum—all under one roof. When health challenges arise, visit thepoultrydoc.com to access a poultry-specific telemedicine platform and get expert guidance on diagnostic and treatment options.

ABOUT THE AUTHOR

NIKKI HUSTED is a former public school teacher who "dropped out of school" to care for her children. While she has a hobby farm that includes garden beds, composting stations, quails, goats, and geese, her primary educational content is focused on chicken keeping. After she started sharing her adventures online, she quickly found a community of like-minded chicken keepers and homesteaders on Instagram, TikTok, Facebook, and YouTube, where she posts as Purely Chickens. She brings additional expertise to the page in her previous book, *Chicken Keeping Pure and Simple*. Find her online as @purelychickens or at her website, purelychickens.com.

INDEX

E

F

H